T0345021

Lateral Solutions to Mathematical Problems

Lateral Solutions to Mathematical Problems offers a fresh approach to mathematical problem solving via lateral thinking. Lateral thinking has long been used informally by good mathematics teachers and lecturers to spice up their material and interest their students in the more artistic aspects of mathematical problem solving. In this book, the author attempts to carry out this process formally, with reference to specific, non-technical problems that are easily understood and explained at an intermediate level.

This book is appropriate for interested high school students, undergraduates and postgraduates, looking for relief from technical materials and also looking for insight into the methodology of mathematics; for teachers and lecturers looking for a novel approach to course material; and anyone interested in both mathematics and lateral thinking.

AK Peters/CRC Recreational Mathematics Series

Series Editors

Robert Fathauer
Snezana Lawrence
Jun Mitani
Colm Mulcahy
Peter Winkler
Carolyn Yackel

Mathematical Puzzles
Peter Winkler

X Marks the Spot
The Lost Inheritance of Mathematics
Richard Garfinkle, David Garfinkle

Luck, Logic, and White Lies
The Mathematics of Games, Second Edition
Jörg Bewersdorff

Mathematics of The Big Four Casino Table Games
Blackjack, Baccarat, Craps, & Roulette
Mark Bollman

Star Origami
The Starrygami™ Galaxy of Modular Origami Stars, Rings and Wreaths
Tung Ken Lam

Mathematical Recreations from the Tournament of the Towns
Andy Liu, Peter Taylor

The Baseball Mysteries
Challenging Puzzles for Logical Detectives
Jerry Butters, Jim Henle

Mathematical Conundrums
Barry R. Clarke

Lateral Solutions to Mathematical Problems
Des MacHale

Basic Gambling Mathematics
The Numbers Behind the Neon, Second Edition
Mark Bollman

For more information about this series please visit: https://www.routledge.com/
AK-PetersCRC-Recreational-Mathematics-Series/book-series/RECMATH

Lateral Solutions to Mathematical Problems

Des MacHale

CRC Press
Taylor & Francis Group
Boca Raton London New York

CRC Press is an imprint of the
Taylor & Francis Group, an **informa** business

AN A K PETERS BOOK

First edition published 2024
by CRC Press
6000 Broken Sound Parkway NW, Suite 300, Boca Raton, FL 33487-2742

and by CRC Press
4 Park Square, Milton Park, Abingdon, Oxon, OX14 4RN

CRC Press is an imprint of Taylor & Francis Group, LLC

© 2024 Des MacHale

Reasonable efforts have been made to publish reliable data and information, but the author and publisher cannot assume responsibility for the validity of all materials or the consequences of their use. The authors and publishers have attempted to trace the copyright holders of all material reproduced in this publication and apologize to copyright holders if permission to publish in this form has not been obtained. If any copyright material has not been acknowledged please write and let us know so we may rectify in any future reprint.

Except as permitted under U.S. Copyright Law, no part of this book may be reprinted, reproduced, transmitted, or utilized in any form by any electronic, mechanical, or other means, now known or hereafter invented, including photocopying, microfilming, and recording, or in any information storage or retrieval system, without written permission from the publishers.

For permission to photocopy or use material electronically from this work, access www.copyright.com or contact the Copyright Clearance Center, Inc. (CCC), 222 Rosewood Drive, Danvers, MA 01923, 978-750-8400. For works that are not available on CCC please contact mpkbookspermissions@tandf.co.uk

Trademark notice: Product or corporate names may be trademarks or registered trademarks and are used only for identification and explanation without intent to infringe.

ISBN: 978-1-032-37699-8 (hbk)
ISBN: 978-1-032-37092-7 (pbk)
ISBN: 978-1-003-34146-8 (ebk)

DOI: 10.1201/9781003341468

Typeset in Epigrafica
by KnowledgeWorks Global Ltd.

*This book is dedicated to my good friend,
mathematician Owen O'Shea,
with respect and admiration.*

Contents

Contents

About the Author

Des MacHale was born in 1946 in Castlebar, County Mayo, Ireland. He earned his B.Sc. and M.Sc. degrees in Mathematical Science at University College Galway. In 1972, he was awarded a Ph.D. degree in Group Theory at the University of Keele in the UK under the supervision of Dr. Hans Liebeck. He taught mathematics for 40 years at the University College Cork in Ireland, rising to the rank of Associate Professor. He has over 100 research articles published in refereed mathematical journals and his interests lie mainly in abstract algebra (groups and rings) but he has also published articles on geometric dissections, number theory, Euclidean geometry, trigonometry and a book on mathematical humour (Comic Sections Plus). Both as a student and teacher, he has won many prizes for his work on the popularization of mathematics. He is currently Emeritus Professor of Mathematics at University College Cork.

In 1984, he set up the Superbrain Examination open to all students of UCC and later the Irish Intervarsity Mathematics Competition. He has written four biographical books on George Boole who was the first professor of mathematics in Cork from 1849 to 1864. These are George Boole His Life and Work (2014), New Light on George Boole (with Yvonne Cohen) (2018), The Poetry of George Boole (2021) and Simply Boole (with Yvonne Cohen) to appear. In 2015, he was awarded an honorary doctorate in Literature by the National University of Ireland for his work on George Boole. He has written over a dozen books of Lateral Thinking puzzles with Paul Sloane and many other books of mathematical puzzles, riddles and jokes – over 80 books in all.

Among Des MacHale's other interests are Geology (in which he has a diploma), puzzles of all sorts, words, music, especially classical and Irish, and photography.

He believes that mathematics is like an infinite goldmine – the more gold you dig out, the more there is to find.

Introduction

There is an old puzzle about a man who is captured by a cruel dictator and is sentenced to death. A scaffold is erected on the seashore where the man is to be hanged, but the dictator offers him one last chance. He gives him an opaque bag containing two pebbles, one black and the other white. The prisoner is allowed to pick one pebble from the bag, sight unseen. If he picks the white pebble, he will be hanged, but if he picks the black pebble, he can go free.

Our hero, rightly suspecting that the dictator is making this offer merely for show and has cheated by putting two white pebbles in the bag, takes out one pebble in his closed hand and throws it far into the sea. Then, he says to the dictator, "If you want to know what colour the pebble I picked was, just look at the colour of the pebble remaining in the bag." This is a wonderful example of using lateral thinking to overcome a seemingly impossible situation. And it is quite mathematical too because it concentrates on the complement of a set, rather than the set itself.

Lateral thinking has been used since ancient times by all the great mathematicians, including Archimedes, Euler, Newton and many others. Archimedes is said to have destroyed the wooden Roman fleet by focusing the sun's rays using mirrors; Euler solved the famous Bridges of Konigsberg problem with a simple lateral parity trick and Newton turned an observation of a falling apple into the magnificent theory of universal gravitation.

Lateral thinking is sideways thinking, slick thinking, smart thinking, often leading to short solutions to difficult problems in mathematics and elsewhere. This book contains 120 mathematical problems and in each case there is a solution based on a lateral twist. Some of the problems are classics but many are new, appearing for the first time. A unique feature of this book is that each solution is followed by "Topics for Investigation," in which the reader is invited to look at problems in a similar vein which follow on from the given problem. This gives rise to hundreds of new problems, some easy, some difficult, but all interesting and exciting. The hope is that the reader, now on the lateral wavelength, will discover lateral solutions to these problems.

Our underlying theme is MIAES, which stands for "Mathematics is an Experimental Science." Many people do not realize that the polished solutions in mathematical textbooks are the result of maybe a dozen failed attempts before near-perfection was achieved. In fact, it is probably true to say that every page of correct and acceptable mathematics is the result of maybe a dozen pages in the recycling bin. If only our teachers had told us that, it could have saved a lot of misery and frustration. It's a bit like those cookery programs on television – nobody believes that the beautiful cakes were the first and only attempt and we suspect that for every successful outcome, there are several burned attempts in the waste bin.

Mathematics is too important a subject for engineering, science, economics, statistics and virtually every organized body of knowledge not to be enjoyed by people, especially pupils and students, worldwide. The secret is first to make the subject interesting and exciting for its own sake, and then useful applications will quickly follow. Yes, the subject can be difficult at times, but it can also be immensely enjoyable and rewarding, especially in the hands of a skilled and enthusiastic teacher. Every mathematician I have met had a great teacher at the second level, and I am no exception. Brother George Sheridan FSC shaped my career in mathematics and gave me an undying love of the subject. He used to compare the subject to a gold mine, which unlike an earthly gold mine, the more gold you found and extracted, the more gold you discovered – a perfect description of mathematics. Yes, the digging is not always easy, but the reward is immense. He also insisted that when I had solved a problem, I had to solve it again in as many different ways as possible. This is a wonderful training for a mathematician.

I honestly believe that lateral thinking is the key to a deep enjoyment of our wonderful subject, and if used properly, this can attract many more people to mathematics. After nearly 60 years of teaching, public lecturing, radio and television appearances and writing books and newspaper articles on mathematics, I have met very many people who have said to me, "Why did nobody ever tell me in school that mathematics could be so enjoyable and that I could understand and do it?"

So enjoy this book, and I hope you get even half as much pleasure out of reading it as I got out of writing it.

Des MacHale, Cork, 2023

PROBLEMS

1 | Number Theory

1.1 If you multiply all the primes less than one million together, what is the final digit of your answer?

1.2 (i) Show that every integer bigger than 5 is the sum of a prime number and a composite number.

(ii) Show that every integer bigger than 22 is the sum of two composite numbers.

1.3 Consider $1! + 2! + 3! + \ldots + n! = \sum_{r=1}^{n} r!$. What is the final digit of this sum? For what values of n is this sum a perfect square?

1.4 Show that it is sensible to define $0!$ to be 1. Can you give a sensible definition for $(-1)!$?

1.5 What is the final digit of 2^{999}?

1.6 Professor Dunderhead has great difficulty in remembering his PIN number for his bank card. All he can recall is that it lies between 5000 and 9999 and that it cannot be expressed as the sum of two or more consecutive positive integers. Can you help the absentminded professor remember his PIN number?

1.7 Can you make up 20 with just three threes and any mathematical operations you like?

1.8 Find all positive integers a and b, with $a^b = b^a$ and $a \neq b$.
Can you deduce which is bigger, e^π or π^e?

1.9 The famous Fibonacci sequence $\{F_n\}$ is defined by $F_1 = F_2 = 1$, and $F_{n+2} = F_n + F_{n+1}$ for all $n > 0$. Thus the sequence is $1, 1, 2, 3, 5, 8, 13, 21, \ldots$
Can you find a nice formula for the sum to n terms of the series

$$1 + 1 + 2 + 3 + 5 + \ldots + F_n = \sum_{r=1}^{n} F_r?$$

1.10 What are the final three digits of 5^{888}?

DOI: 10.1201/9781003341468-1

3

2 **Primes and Divisibility**

2.1 If $f(n) = n^4 + 4$, notice that $f(1) = 5$ is a prime number. For what other positive integer values of n is $f(n)$ a prime number?

2.2 If n is a positive integer, show that $(n!)^2$ divides $(2n)!$

2.3 Show that for each positive integer n, $2n^3 + 3n^2 + n$ is an integer multiple of 6.

2.4 For what positive integer values of n is $n^5 + n + 1$ a prime number?

2.5 Which integer n less than 100 has exactly 7 different divisors (including itself)?

2.6 Show that the positive integers $10n + 1$, $10n + 11$ and $10n + 21$ cannot all be prime numbers.

2.7 Write down any number of the form $ababab$, where a and b are digits – for example 383838. Show that any such number is divisible by 3, 7, 13 and 37.

2.8 Consider the sequence of prime numbers
$\{P_n\} = 2, 3, 5, 7, 11, 13, 17, 19, 23, \ldots$, where P_n is the nth prime. A beautiful known inequality states that $P_{n+2} \leq P_n + P_{n+1}$.
Find all values of n for which the equality $P_{n+2} = P_n + P_{n+1}$ holds.

2.9 A famous unsolved problem in mathematics is the Goldbach conjecture. This states that any even integer greater than 2 is the sum of two prime numbers. Prove that there are infinitely many even numbers for which the Goldbach conjecture is true.

2.10 If $\{P_r\}$ is the sequence of prime numbers, show that

$$\sum_{r=1}^{n} \frac{1}{P_r}$$

is never an integer for any positive integer n.

DOI: 10.1201/9781003341468-2

3 Geometry

3.1

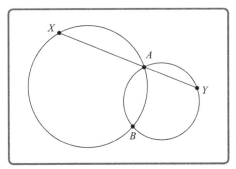

Two circles in the plane intersect at points A and B. What is the maximum possible length of the line segment XAY and in what position?

3.2 Given a semicircular piece of glass, show how to cut it into four pieces which can be reassembled to form an "elliptical" tabletop.

3.3 Each of three circles in the plane intersects the other two in exactly two points, giving exactly six distinct intersection points in all. Show that the common chords of each two circles are concurrent, that is, meet at a single point.

3.4 Show how to cut a cube to form a regular hexagon.

3.5 If L is a line in the plane with equation $ax + by + c = 0$, and p is a point with coordinates (x_1, y_1), find the shortest distance from p to L.

3.6 Let S be a finite set of points in the plane such that every line which contains two points of S also contains a third point of S. Show that all points of S lie on the same straight line.

DOI: 10.1201/9781003341468-3

3.7

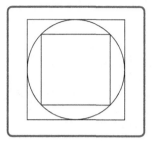

If the inner square has an area of 7 square units, what is the area of the outer square?

3.8 Show that the sum of the angles of a plane triangle is 180°.

3.9 The famous number $\phi = (1/2)(\sqrt{5} + 1)$ is called the golden ratio and is usually associated with a regular pentagon. Show how to construct ϕ using just an equilateral triangle.

3.10 How many observations of the position of a planet or a moon must be made so that the complete path of the planet or moon can be determined?

4 Trigonometry

4.1 Evaluate $[(1 + \tan 15°)/(1 - \tan 15°)]^2$ with as little calculation as possible.

4.2 Find all triangles ABC in which all of $\tan A$, $\tan B$ and $\tan C$ are positive integers.

4.3 Find the sum of the infinite series

$$\sum_{n=1}^{\infty} \sin^{-1}\left(\frac{1}{\sqrt{n^4 + 2n^3 + 3n^2 + 2n + 2}}\right)$$

4.4

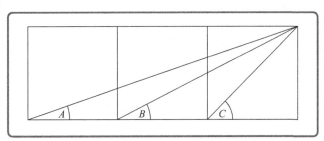

Three squares lie side by side. Show that $\angle C = \angle A + \angle B$.

4.5 In a triangle ABC, $a = 6$, $b = 5$ and $c = 4$. Show that the angle A is twice the size of $\angle C$.

4.6 Show that $3 < \pi < 4$.

4.7 Find a formula for the area of a triangle involving only two angles and the length of just one side.

DOI: 10.1201/9781003341468-4

4.8

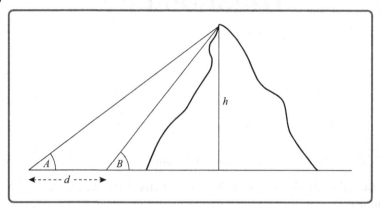

Find a formula for the height of the mountain in terms of the $\angle A$ and $\angle B$ and the distance d.

4.9 Use Euler's identity $e^{i\theta} = \cos\theta + i\sin\theta$, where $i^2 = -1$, to show that $\sin^2\theta + \cos^2\theta = 1$.

4.10 If two internal angle bisectors of a triangle have the same length, prove that the triangle is isosceles.

5   Probability

5.1 Given three distinct points on the surface of a sphere, what is the probability that there is a hemisphere on which they all lie?

5.2 The length and breadth of a rectangle are both a whole number of feet. What is the probability that the area of the rectangle is an odd number of square feet?

5.3 *A* and *B* play a game in which each of them tosses their own coin simultaneously. Beforehand, each of them nominates a triple such as (heads)(tails)(heads) and the winner is the person whose nominated triple occurs first. Devise a strategy for winning this game.

5.4 A thin rod is broken at random into three pieces. What is the probability that these three pieces can be used to form a triangle?

5.5 Three cubical dice *A*, *B* and *C* have the following numbers on their faces:

A	2,	2,	4,	4,	9,	9
B	1,	1,	6,	6,	8,	8
C	3,	3,	5,	5,	7,	7

If *A* is rolled against *B*, what is the probability that *A* will win? If *B* is rolled against *C*, what is the probability that *B* will win? Finally, if *C* is rolled against *A*, what is the probability that *C* will win?
Can you explain what is happening?

5.6 A coin is tossed three times. What is the probability that the outcome will be two heads and one tail?

5.7 What is the probability that your father and mother have birthdays six months or less apart?

5.8 A random point is picked within a circular disc of radius two units. What is the probability that the point is nearer to the center of the disc than to the circumference?

DOI: 10.1201/9781003341468-5

5.9 A random chord is drawn in a circle. What is the probability that the chord is longer than the side of an equilateral triangle inscribed in the circle?

5.10 There are four children in a family. What is the probability that these children will consist of three girls and one boy?

6 Combinatorics

6.1 There are six people at a party. Show that there are three people, each of whom knows the other two, or that there are three people, none of whom knows either of the other two.

6.2 How many positive integers are there which do not repeat a digit?

6.3 You go into a darkened room and you open a drawer containing four identical red socks, five identical blue socks and six identical green socks. What is the minimum number of socks you must withdraw from the drawer to ensure a matching pair of socks? What is the minimum number of socks you must withdraw from the drawer if you want a pair of blue socks?

6.4

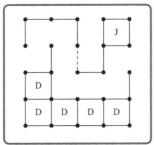

It is not often one gets a chance of doing this! This is a puzzle taken from a book by one of the world's most famous (and best!) mathematical problem setters. It concerns a game of dots and boxes played by David and Jane. The problem asks you to show that if David moves by drawing the dotted line, he must win the game. Show how Jane can think laterally and legally win the game.

6.5

4	9	2
3	5	7
8	1	6

DOI: 10.1201/9781003341468-6

13

Many people know about <u>additive</u> magic squares – a 3 × 3 array of nine different numbers such that all rows, columns and length three diagonals add to the same total. But can you form a 3 × 3 <u>multiplicative magic square</u>, where all rows, columns and length three diagonals multiply to the same integer? All entries of course must be different integers.

6.6 Twelve different books are placed in a row on a shelf. In how many different ways can this be done if one particular book A must never be adjacent to another particular book B?

6.7

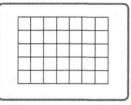

A chocolate bar has 35 squares in a 7 × 5 block. It can be snapped only along the vertical and horizontal lines, and you are not allowed to stack pieces after snapping. What is the minimum number of snaps needed to produce the 35 small squares of chocolate?

6.8

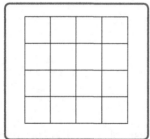

Place the integers 1 to 16, one in each box so that all rows, columns and length four diagonals add up to <u>different</u> totals (an anti-magic square), with the smallest total 29 and the largest total 38.

6.9 Show that a set with precisely n elements has exactly 2^n subsets, where n is a positive integer.

6.10 How many different divisors does $1^1 2^2 3^3 4^4 5^5 6^6$ have?

Dissections

7.1 Show how to dissect a square into nine pieces, which can then be reassembled to form five equal squares.

7.2

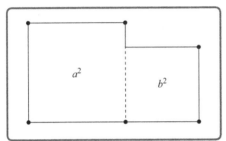

Consider squares of sides a and b where $a > b$. Attach squares of these sidelengths together to form an L-shaped figure. Show how to dissect this figure into three different pieces which can be reassembled to form a square of side c, where $a^2 + b^2 = c^2$.

7.3

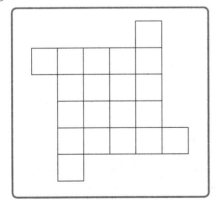

Show how to cut this figure into four pieces which can be reassembled to form a square.

DOI: 10.1201/9781003341468-7

7.4 Show how to dissect an equilateral triangle into four pieces which can be reassembled to form two equilateral triangles.

7.5 Show that a cube cannot be dissected into a finite number (≥ 2) of cubes, no two of which have the same size.

7.6 What is the minimum number of pieces into which a 30×30 square piece of carpet must be cut so as to cover a 25×36 floor exactly?

7.7

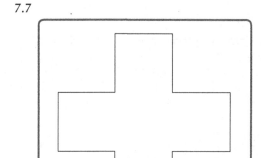

Can you cut this shape into four pieces and reassemble them to form a square?

7.8 Show how to "prove" the famous theorem of Pythagoras by a dissection method.

7.9

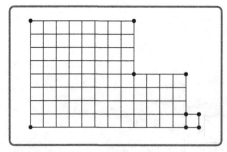

Show how to cut this figure, consisting of squares of sides 8, 4 and 1 attached together, into three pieces which can be reassembled to form a square.

7.10

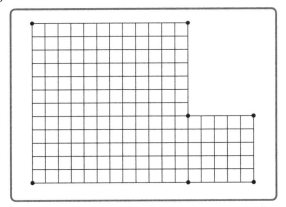

Show how to dissect this figure, a square of side 12 attached to a square of side 5, into three pieces which can be reassembled to form a square. [You may cut only along the marked lines.]

Matchsticks and Coins

8.1

Move just one matchstick to make the equation correct.

8.2

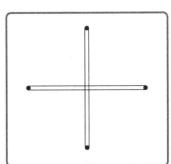

Move just one matchstick to make a square.

8.3

Move just two matchsticks to make the equation valid.

8.4

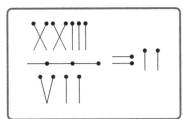

Move just one matchstick to show a well-known approximation.

DOI: 10.1201/9781003341468-8

8.5

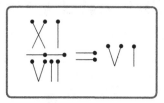

Move just one matchstick to make this equation valid.

8.6

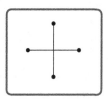

Add just one matchstick to make a square.

8.7

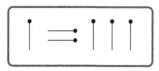

Make a correct equation by moving just two matchsticks. There are several different solutions, so how many can you find?

8.8

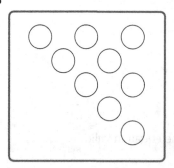

Move two coins to turn the triangle into a square.

8.9

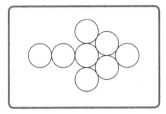

Move two coins to alter the direction of the arrow.

8.10

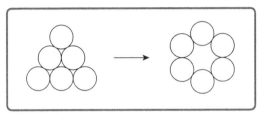

Change the first configuration of coins to the second configuration of coins in four moves. After each coin is moved, it must touch at least two other coins.

9 | Logic

9.1 An old puzzle goes as follows: A man is looking at a painting and he says, "Brothers and sisters I have none, but this man's father is my father's son." Whose painting is he looking at?
The classical solution is that he is looking at a painting of his son.
Can you think laterally and produce another valid solution?

9.2 Is "no" the answer to this question?

9.3 1. This list contains exactly one false statement.

   2. This list contains exactly two false statements.

   3. This list contains exactly three false statements.

   4. This list contains exactly four false statements.

   5. This list contains exactly five false statements.

   Which of the above statements are true and which are false?

9.4 A lady sees a pair of beautiful boots in a shop window with a price tag of $30. She goes into the shop and buys the boots, for which she pays a $50 bill. The shoe shop man has no change, so he goes into the shop next door and exchanges the $50 bill for two $20 bills and a $10 bill. He gives $20 change to the lady and puts $30 in the till. The lady departs.
A few minutes later, the shopkeeper next door storms in and says that the $50 bill he has just changed is a forgery and demands his money back. Our hero is forced to give him $50 from the till.
The question is, how much did the shoe shop man lose in the transaction?

9.5 A man with a fox, a duck and a bag of corn wishes to convey them across a river in a small boat which can safely carry only himself and one of his three passengers. The fox cannot be left alone with the duck and the duck cannot be left alone with the bag of corn, on either side of the river, at any time. Almost everyone knows the classical answer to this problem, which requires a minimum of seven crossings (duck across first, return for fox, fox across, bring duck back, bring corn across, back for duck, duck across) but can you come up with a lateral solution which requires just three crossings?

DOI: 10.1201/9781003341468-9

9.6 You are given two ropes, each of which takes exactly 60 minutes to burn. They are made of different materials and they burn at different rates and inconsistently. By burning the ropes, how can you measure 45 minutes exactly?

9.7 A has three bottles of wine and B has five bottles of wine; C has no bottles of wine. C pays A and B $8 to share equally in their drinking session. How should A and B share the money between them?

9.8 How many letters are there in the correct answer to this problem?

9.9 There are two tribes on an island – members of one tribe always tell the truth, while members of the other tribe always lie, but I cannot tell the two tribes apart. I come across one of them, at a fork in the road; one branch leads to the city, the other does not. By asking him just one question, how can I find the correct road to the city?

9.10 I am captured by a crowd of savages who condemn me to death. The chief, who has a doctorate in logic, says that I am allowed to make one statement. If the statement is true, I will be shot, but if the statement is false, I will be hanged. What statement should I make?

10 Maxima and Minima

10.1 A set of positive integers has sum 100. What is the maximum value their product can have?

10.2 A piece of wire of length *l* is bent into the shape of a sector of a circle. What is the maximum area the sector can have? (A non-calculus solution is possible and more lateral.)

10.3 Two sides of an isosceles triangle have length 10 cm each. What length of the third side will maximize the area of the triangle?

10.4 What is the largest number that can be written with just three digits and no other symbols or operations?

10.5 If *x* is a positive real number, what is the minimum value that $x + (1/x)$ can have? (Again there is a nice lateral solution without calculus.)

10.6

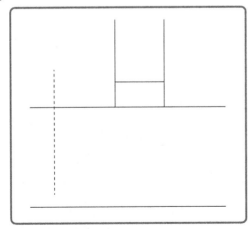

A football must be placed on the dotted line and kicked toward the goal posts. How can one find the optimal point on the dotted line on which to place the football?

10.7 If $a > 0$, and *b* and *c* are real numbers, find the minimum value of $ax^2 + bx + c$ as simply and as laterally as possible.

DOI: 10.1201/9781003341468-10

10.8 What is the maximum and minimum number of Friday the 13ths that can occur in any calendar year?

10.9 A right circular cylinder, of radius r and height h, is closed at both ends. If the total surface area A of the cylinder is constant, what value of the ratio h/r ensures that the cylinder has a maximum volume V?

10.10 What is the biggest integer, written in the English language, where the letters are in alphabetical order?

Calculus and Analysis

11.1 Find the limit of $(2^n/n!)$ as n tends to infinity. (Be as lateral as possible.)

11.2 Find the equation of the tangent to the curve $y = x^2$ at the point $(3, 9)$ without using differentiation or limits.

11.3 Evaluate $\int \sec x\, dx$

11.4 If $x_1, x_2, \ldots, x_n$ are positive real numbers, prove the Arithmetic Mean–Geometric Mean (AM-GM) inequality

$$(1/n)(x_1 + x_2 + \ldots + x_n) \geq \sqrt[n]{x_1 x_2 \cdots x_n}$$

11.5 Which of the following numbers is bigger and why?

$$\int_0^1 \sqrt[4]{1 - x^7}\, dx \quad \text{and} \quad \int_0^1 \sqrt[7]{1 - x^4}\, dx.$$

11.6 Evaluate

$$\int_{\pi/4}^{\pi/2} \left(\frac{\sin x}{x}\right)^2 dx - \int_{\pi/4}^{\pi/2} \left(\frac{\sin 2y}{y}\right) dy$$

11.7 Find the 20th derivative $\frac{d^{20}y}{dx^{20}}$ of the function $y = 1/(1 - x^2)$.

11.8 Evaluate

$$\int \frac{dx}{\sin x + \cos x}.$$

11.9 If e^x is the exponential function and $e^a = e^b$, show that $a = b$.

11.10 Construct a vessel which has infinite surface area, but a finite volume.

DOI: 10.1201/9781003341468-11

12 A Mixed Bag

12.1 How I wish I could remember x easily today. What is the value of x?

12.2 If lines l_1 and l_2 are perpendicular to line l_3, give an <u>algebraic</u> proof that l_1 is parallel to l_2.

12.3 If A and B are $n \times n$ matrices with $AB = I_n$, the identity $n \times n$ matrix, show that $BA = I_n$ also.

12.4 Express $x^4 + y^4 + z^4 + w^4 - 4xyzw$ as the sum of four squares. What can you deduce?

12.5 If $ax^2 + bx + c = 0$ and $a \neq 0$, solve the equation without completing the square.

12.6 By considering

$$\int_0^1 \frac{x^4(1-x)^4}{1+x^2}\,dx$$

show that $\pi < 22/7$.

12.7 Undoubtedly $\sqrt{5} - 2 = \sqrt{5} - 2$.
By adding three strokes to the right-hand side of this equation, preserve the validity of the equation.

12.8 A bar of soap can be made from seven soap ends. How many bars of soap can be made from 49 soap ends?

12.9 Find all terms of the sequence 1, 11, 111, 1111, 11111, ... which are perfect squares, where the nth term consists of n decimal digit ones.

12.10 A very tough puzzle to finish.
It is known that a square can be dissected into a finite number (>1) of different squares, but that an equilateral triangle cannot be dissected into a finite number (>1) of different equilateral triangles. Can an isosceles right-angled triangle be dissected into a finite number (>1) of different isosceles right-angled triangles?

DOI: 10.1201/9781003341468-12

SOLUTIONS

1 Number Theory

1.1 If you have a weekend to spare and a fast computer, you may of course be able to get the actual answer, but think laterally!

In writing out the product, you get $2, 3, 5, 7, \ldots$; notice that there is a 2 in that there and a 5, so the answer has 10 as a factor, so the final digit must be 0.

Topic for Investigation: What can you say about the penultimate digit?

1.2 These questions arise naturally from the famous Goldbach conjecture (unsolved at the time of writing, though progress has been made). This states that every even number greater than 2 is the sum of two prime numbers – fame and fortune await you if you can solve this innocent-looking problem.

(i) We note by simple experiment that none of the numbers $1, 2, 3, 4, 5$ is the sum of a prime number and a composite number.

So, we can assume $n > 5$. Now a very clever lateral technique for solving mathematical problems is to divide and conquer, that is, split the problem into several cases and solve the problems in each case, sometimes using different techniques. Here, let $n > 5$ and consider the cases where n is odd and n is even.

If n is odd, then $n = 2k + 1 = 3 + 2k - 2 = 3 + 2(k - 1)$ for some integer k, which is the sum of the prime 3 and the composite $2(k - 1)$. Note that since $n > 5$, $k > 2$, so $2(k - 1)$ is properly composite.

If n is even, then for some integer k, $n = 2k = 2k - 2 + 2 = 2(k - 1) + 2$, which is the sum of the prime 2 and the composite $2(k - 1)$. Again since $n > 5$, $k > 3$, so $k - 1 > 1$.

Of course the expression of a number as the sum of a prime and a composite can be achieved in general in several different ways. For example, $20 = 2 + 18 = 5 + 15 = 11 + 9$.

(ii) We notice that $8 = 4 + 4$ and $10 = 4 + 6$ can both be expressed as the sum of two composites, but a little experimentation will show that $1, 2, 3, 4, 5, 6, 7, 9$, and 11 cannot be so expressed. So suppose $n > 11$.

DOI: 10.1201/9781003341468-13

Again we use divide and conquer.

If $n = 2k$ is even, then $2k = 2k - 4 + 4 = 2(k - 2) + 4$, which is the sum of two composites, since $k - 2 > 1$ as $n > 11$.

If $n = 2k + 1$, then $n = 2k - 8 + 9 = 2(k - 4) + 9$, which is the sum of two composites as $k > 4$.

Again we notice that writing an integer as the sum of two composites can be achieved in general in many different ways. For example, $16 = 4 + 12 = 6 + 10 = 8 + 8$.

TOPIC FOR INVESTIGATION: For a given n, in how many different ways can n be expressed as the sum of a prime and a composite, and the sum of two composites? Can you see any pattern? Read about the Goldbach conjecture.

1.3 Let $S_n = \sum_{r=1}^{n} r! = 1! + 2! + \ldots + n!$
We see that $S_1 = 1, S_2 = 3, S_3 = 9, S_4 = 33$.

For $n > 4$, we notice that $n!$ is a multiple of 10, so that S_n always has the final digit 3. So the answer is 1 for $n = 1$, 9 for $n = 3$ and 3 for every other positive integer.

When is S_n a square? Clearly for $n = 1$ and 3, S_n is a square. For all other n, S_n ends in 3 and cannot be a square, since a square must end in $0, 1, 4, 5, 6,$ or 9. To see this, look at $(10i + j)^2$ for $0 \le j \le 9$.

TOPICS FOR INVESTIGATION: Nobody seems to know a closed formula for $S_n = \sum_{r=1}^{n} r!$. Perhaps you can be the first person to find one! What sort of function would it have to be? Remember that $S_n - S_{n-1} = n!$. Can you use that?

Can you spot a number that always divides S_n for $n > 1$? For $n > 9$? When does S_i divide S_j?

1.4 For $n = 1$, we define the factorial function $n!$ by $1! = 1$.
Assume that $k!$ has been defined for some k in $\mathbb{N}$.
Then we define $(k + 1)!$ to be $(k + 1)(k!)$.
The Axiom of Induction now assures us that $n!$ is defined for all $n \in \mathbb{N}$.
So $1! = 1; 2! = (2)(1!) = 2; 3! = (3)(2!) = (3)(2) = 6; 4! = (4)(3!) = (4)(6) = 24; (5!) = (5)(4!) = (5)(24) = 120$, and so on. Thus we often write $n! = (n)(n - 1) \ldots \cdot 3 \cdot 2 \cdot 1$. How would one define 0!? Well, it is up to us, but we would like 0! to satisfy $(0 + 1)! = (0 + 1)(0!)$ to preserve the symbolic form above. Thus we would like $1 = 1! = (1)(0!) = 0!$, so it would make sense to define 0! to be 1, and most mathematicians agree to do this; but it is just a definition!
Another motivation comes from combinatorics – the number of ways of combining n objects, r at a time, when $n > r$ is $n!/r!(n - r)!$. This is often written ^nC_r, $\binom{n}{r}$ or $C(n, r)$.

What happens if we let $n = r$? A priori, the formula makes no sense, but suppose we just try it.

We get

$$^nC_n = \frac{n!}{n!\,0!}.$$

But there is just one way of combining n things n at a time so $^nC_n = 1 = \frac{n!}{n!0!}$. This again indicates that it would be desirable to let $0! = 1$. What about a sensible definition for $(-1)!$? Well, preserving $(n + 1)! = (n + 1)n!$, we would get for $n = -1$, $0! = 0(-1)!$. Now we have agreed that $0! = 1$, so this becomes $1 = 0(-1)!$

We have a dilemma: if $(-1)!$ exists, then $1 = 0$, which would wreck the rest of mathematics! The only way out is to decide that $(-1)!$ does not exist or in other words, $(-1)!$ cannot be sensibly defined!

The same problem arises if we try to define $(-2)!, (-3)!, \ldots, (-n)!, \ldots$ for $n \in \mathbb{N}$. It is wiser to agree that these are undefined.

TOPIC FOR INVESTIGATION: Would it be possible to define $\left(\frac{1}{2}\right)!$ or $(\sqrt{2})!$ or $\left(\frac{10}{3}\right)$? It turns out that these numbers can be defined and are very useful in Probability and Statistics.

Read about the Gamma Function – a beautiful concept invented by the great mathematician Euler (1707–1783).

1.5 Look at the first few values of 2^n.

$$2^1 = 2; 2^2 = 4; 2^3 = 8; 2^4 = 16; 2^5 = 32; 2^6 = 64; \ldots$$

Notice the repeating pattern of end digit

$$2, 4, 8, 6, 2, 4, 8, 6, 2, 4, 8, 6, \ldots$$

It is easy to see that we get a cycle of length 4:
2^{4k} ends in 2; 2^{4k+1} ends in 4; 2^{4k+2} ends in 8; and 2^{4k+3} ends in 6. $999 = 4 \cdot 249 + 3$ so 2^{999} has the final digit 6.

TOPIC FOR INVESTIGATION: What happens if we replace 2 by 3, for example?

1.6 One of the best lateral tricks in mathematics is knowledge – the more mathematical results you know, the more advantage you have over problems. For example, did you know the following beautiful result?

A positive integer is the sum of two or more consecutive positive integers if and only if it is not a power of 2.

Thus $3 = 1 + 2$, $5 = 2 + 3$, $14 = 2 + 3 + 4 + 5$, etc. However, $2, 4, 8, 16, 32, \ldots$ cannot be expressed as the sum of two or more consecutive positive integers.

Meanwhile, back at the problem of the Professor's PIN, it turns out that there is only one power of 2 between 5000 and 9999 and that is $2^{13} = 8192$. So that is the number he wants to remember.

TOPIC FOR INVESTIGATION: Find a proof of the theorem used above – that a positive integer is the sum of two or more consecutive positive integers if and only if it is not a power of 2. There are lots of proofs in textbooks, problem books, and on the web, but remember, a proof you find yourself is much more valuable and less easily forgotten.

Also, let $f(n)$ be the total number of different ways that the positive integer n can be expressed as the sum of consecutive integers. What information can you find about $f(n)$?

1.7 There may be many ways of making 20 with just three 3s, but here is one way:

$$20 = \frac{(3 + 3)}{.3}$$

TOPICS FOR DISCUSSION: What other ways can you find? What other numbers can be made with three 3s? In particular, can you make 64 in this way?

1.8

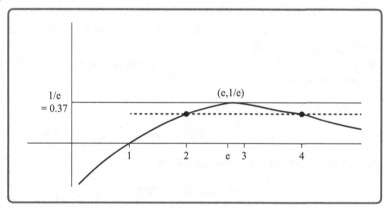

When a solution in positive integers is sought for an equation, a lateral technique often used is to consider rational or even real solutions of the equation.

Since $a^b = b^a$, we have $b \ln a = a \ln b$, where ln is log to the base e. Thus

$$\frac{\ln a}{a} = \frac{\ln b}{b}$$

and we are moved in the direction of considering the function

$$f(x) = \frac{\ln x}{x}$$

for positive real x. The graph of the function looks something like this – it is increasing in the interval $(0, e)$ and decreasing for $x > e$. By ordinary calculus, it has a maximum value of $\frac{1}{e}$ at $x = e$.

Our only hope of $\frac{\ln a}{a} = \frac{\ln b}{b}$ for positive integers a and b is by drawing horizontal lines through $(2, f(2))$ and $(1, f(1))$.

Since $f(x)$ is never zero for $x > 1$, the second possibility does not work. Luckily $f(2) = f(4)$ so $\frac{\ln 2}{2} = \frac{\ln 4}{4} \left(= \frac{2\ln 2}{2 \cdot 2} \right)$ is the only solution giving $2^4 = 4^2$ as the only integer solution of the original equation $a^b = b^a$, where $a \neq b$. Since $\frac{\ln e}{e}$ is the maximum value of $f(x)$ attained at $x = e$ and $\pi > e$, we see that $\frac{\ln e}{e} > \frac{\ln \pi}{\pi}$, $\pi \ln e > e \ln \pi$, and so $e^\pi > \pi^e$.

TOPIC FOR DISCUSSION: What are the rational number solutions of the equation $a^b = b^a$, where $a \neq b$?

Here is one solution: $a = 9/4, b = 27/8$;

Can you find others?

All the others?

1.9 Rewrite the equation $F_{n+2} = F_n + F_{n+1}$ as

$$
\begin{array}{rcl}
F_{n+2} - \cancel{F_{n+1}} &=& F_n \\
\cancel{F_{n+1}} - \cancel{F_n} &=& F_{n-1} \\
\cancel{F_n} - \cancel{F_{n-1}} &=& F_{n-2} \\
\cancel{F_{n-1}} - \cancel{F_{n-2}} &=& F_{n-3} \\
\vdots \quad \vdots && \vdots \\
\cancel{F_5} - \cancel{F_4} &=& F_3 \\
\cancel{F_4} - \cancel{F_3} &=& F_2 \\
\cancel{F_3} - F_2 &=& F_1 \\
\hline
F_{n+2} - F_2 &=& \sum_{i=1}^{n} F_i
\end{array}
$$

adding both sides and canceling. So $\sum_{i=1}^{n} F_i = F_{n+2} - 1$. But the story does not end here – how can we find the value of F_{n+2}?

Astonishingly, it turns out that

$$F_n = \frac{(1 + \sqrt{5})^n - (1 - \sqrt{5})^n}{2^n \sqrt{5}}$$

so that

$$\sum_{i=1}^{n} F_i = \frac{(1 + \sqrt{5})^{n+2} - (1 - \sqrt{5})^{n+2}}{2^{n+2} \sqrt{5}} - 1$$

and there is no way of doing this without recourse to $\sqrt{5}$!

1.10 Again we look for a pattern:

$$5^1 = 5; \qquad 5^2 = 25; \qquad 5^3 = 125; \qquad 5^4 = 625;$$
$$5^5 = 3125; \qquad 5^6 = 15625; \qquad 5^7 = 78125$$

We see that, except for the first two cases, 5^{2n-1} ends in 125 and 5^{2n} ends in 625.

Thus 5^{888} ends in 625.

2 Primes and Divisibility

2.1 The polynomial $f(n) = n^4 + 4$ does not look as if it has factors, but it does! Now here is the lateral bit: write $n^4 + 4$ as $n^4 + 4n^2 + 4 - 4n^2 = (n^2 + 2)^2 - (2n)^2 = (n^2 + 2n + 2)(n^2 - 2n + 2)$.

If this number is to be a prime, the smaller factor must be 1, so we get $n^2 - 2n + 2 = 1$, so $n^2 - 2n + 1 = 0$, or $(n-1)^2 = 0$. Thus $n = 1$, $f(1) = 5$ and so 5 is the only prime of the form $n^4 + 4$.

TOPICS FOR INVESTIGATION: What primes have the form $n^4 + n^2 + 1$, where n is a positive integer?
What primes have the form $x^4 + 4y^4$, where x and y are positive integers?
What primes have the form $x^4 + x^2y^2 + y^4$, where x and y are positive integers?

2.2 "Think outside the box" is a phrase often used to describe lateral thinking, but one can go further – "don't just think outside the box, think in another box altogether"!

How about mathematical induction? Try it and see – tricky and not very pleasant! How about combinatorics? Suppose you had six objects, the first three of which are indistinguishable from each other and the second three indistinguishable from each other too, though different from the first three. How many different arrangements of the six objects are possible?

Well, there are 6! arrangements of the six objects but these contain some repetitions. The first three objects can be rearranged in 3! different ways as can the second three objects. Thus there are 6!/3!3! distinct arrangements. But by definition, this number must be an integer, so $(3!)^2$ must divide 6!.

There is nothing particular about 3 and 6 here – the reasoning is perfectly general for $2n$ objects split into two groups of n objects each and so we can conclude that $(n!)^2$ divides $(2n)!$ for each n.

TOPIC FOR INVESTIGATION: Can you generalize this result? Suppose that $a_1 + a_2 + \ldots + a_r \leq n$. What can you say about $(a_1)!(a_2)! \ldots (a_r)!$?

2.3 This problem could certainly be solved directly by mathematical induction or several other techniques, but here is a beautiful lateral solution. Thinking in another box, we look at the summation of series.

DOI: 10.1201/9781003341468-14

It is well known that the sum of the first n integer squares is $(n/6)(n + 1)(2n + 1)$, that is,

$$S_n = \sum_{r=1}^{n} r^2 = (n/6)(n + 1)(2n + 1),$$

so $6S_n = 2n^3 + 3n^2 + n$.

Since S_n is clearly an integer, we see at once that $2n^3 + 3n^2 + n$ is an integer multiple of 6.

<u>Topic for Investigation</u>: Find the values of $\sum_{r=1}^{n} r^3$ and $\sum_{r=1}^{n} r^4$ and in each case deduce an integer divisibility result.

2.4 If $n = 1$, then $n^5 + n + 1 = 3$ is certainly prime, so the phenomenon does arise. Now $n^5 + n + 1$ does not appear to have factors, but it does! But how is one to find them?

We move out of the box of prime numbers into the apparently unrelated box of complex numbers. Recall that if ω is a complex number but not real cube root of unity, then $\omega \neq 1$ but $\omega^3 = 1$ and $1 + \omega + \omega^2 = 0$. [Actually $\omega = (1/2)(-1 \pm \sqrt{-3})$, but we do not actually need to know that fact here.]

Consider $\omega^5 + \omega + 1$. Since $\omega^3 = 1$, we have $\omega^5 + \omega + 1 = \omega^2 + \omega + 1 = 0$. This implies that $(n - \omega)$ is a factor of $n^5 + n + 1$. Similarly, $(\omega^2)^5 + \omega^2 + 1 = \omega^{10} + \omega^2 + 1 = \omega^9 \cdot \omega + \omega^2 + 1 = \omega + \omega^2 + 1 = 0$. So $(n - \omega^2)$ is also a factor of $n^5 + n + 1$ and thus $(n - \omega)(n - \omega^2) = n^2 - (\omega + \omega^2)n + \omega^3 = n^2 + n + 1$ is a factor of $n^5 + n + 1$.

By long division of polynomials (remember that?) or otherwise, we get

$$n^5 + n + 1 = (n^2 + n + 1)(n^3 - n^2 + 1).$$

If $n^5 + n + 1$ is to be a prime number then either:

(i) $n^2 + n + 1 = 1$, so $n(n + 1) = 0$, implying $n = 0$ or $n = -1$, which is not possible, or

(ii) $n^3 - n^2 + 1 = 1$, so $n^3 - n^2 = 0 = n^2(n - 1)$. So $n = 0$, which is not possible, or $n = 1$. Thus $n = 1$ is the only possibility, and we have already seen that this gives rise to the prime 3.

<u>Topic for Investigation</u>: For what other values of k can you find all the primes of the form $n^k + n + 1$?

2.5 A piece of good advice when solving mathematical problems is always to experiment first with a specific case before tackling the general problem.

Consider the divisors of 24:

$$1, 2, 3, 4, 6, 8, 12, 24.$$

Notice that each divisor d gives rise to a corresponding divisor $24/d$. Here the divisors occur in distinct pairs so that there are an even number of them. In fact, there is only one situation where the number of divisors is odd and that is when we deal with a square number. Consider, for example, $n = 81$. Its divisors are 1, 3, 9, 27, 81 and in general $d = n/d$ because $n = d^2$.

Since we seek a number less than 100 with exactly 7, which is an odd number, of divisors, we need to look only at squares. Consider p^2, where p is a prime number. This has just 1, p, p^2 as divisors so we can eliminate it. Similarly, we can eliminate p^4, which has exactly five divisors 1, p, p^2, p^3, p^4.

Thus we are left with 36 and 64. We see that 36 has 9 divisors 1, 2, 3, 4, 6, 9, 12, 18 and 36 so we are left with the unique answer 64 which has exactly 7 divisors 1, 2, 4, 8, 16, 32 and 64.

Topic for Investigation: Given an integer k, when can we find an integer n^2 with exactly $2k + 1$ distinct divisors?

2.6 One of the greatest lateral techniques in mathematics was discovered by Gauss (1777–1855). It is called congruence of integers. A simple example is that when an integer is divided by 3, it leaves as remainder exactly one of 0, 1 or 2. Thus every integer is of the form $3k$ or $3k + 1$ or $3k + 2$, for some integer k.

Suppose now that for some n, $10n + 1$, $10n + 11$ and $10n + 21$ are all prime numbers.

If $n = 3k$, then $10n + 21 = 30k + 21 = 3(10k + 7)$, which is not a prime number.

If $n = (3k + 1)$, then $10n + 11 = 10(3k + 1) + 11 = 30k + 21 = 3(10k + 7)$, which is not a prime number.

If $n = 3k + 2$, then $10n + 1 = 10(3k + 2) + 1 = 30k + 21 = 3(10k + 7)$, which is not a prime number.

Topic for Investigation: What other non-trivial triples of integers can you discover that cannot all be prime numbers?

2.7 First consider an example, say, 383838. Now

$$383838 = 380000 + 3800 + 38 \quad = 38(10000 + 100 + 1) = (38)(10101).$$

Now it is not difficult to see that $10101 = 3 \times 7 \times 13 \times 37$ and the result follows in this case.

In general, with a slight abuse of notation, $ababab = (ab)(10101)$ is divisible by $10101 = 3 \times 7 \times 13 \times 37$, and when we divide by all of these numbers successively, we are left with the digit pair we started with.

This can be the basis of a very impressive calculator trick. Ask somebody to key in a number of the form $ababab$.

Now ask the victim to divide the number by 3 – what, no remainder?

Now, seemingly picking the numbers at random, ask the victim to divide the answer by 7, 13 and 37 and bet the answer will be the same as the pair of digits they started with. Magic! Emphasize that you did not know what pair they started with – of course it does not matter and the trick can be performed with a whole class simultaneously.

TOPICS FOR INVESTIGATION:

(i) Show that the numbers of the digit structure $abcabc$ are divisible by 7, 11 and 13. Show how to use this as a magic trick.

(ii) Show that numbers of the digit structure $abcdabcd$ are divisible by 137. What is the "magic" trick this time?

(iii) What divisors can you guarantee a number of the digit structure $abababab$ will always have?

2.8 This problem looks hard, but it is in fact a lateral "quickie."

If $P_n > 2$, then P_n, P_{n+1} and P_{n+2} are all odd, so $P_n + P_{n+1}$ is even.

Thus $P_{n+2} = P_n + P_{n+1}$ is not possible.

So P_n must be 2 so $5 = 2 + 3$ is the only solution.

TOPIC TO INVESTIGATE: What can you say about the sequence $\{a_n\}$ given by $a_n = P_n + P_{n+1} - P_{n+2}$?

2.9 Another lateral quickie, just to be unpredictable!

There are infinitely many distinct prime numbers p, and $2p$ is always even. So we have produced infinitely many even numbers and for each of these $2p = p + p$ is the sum of two prime numbers. Unfortunately, infinitely many does not mean all!

TOPICS FOR INVESTIGATION:

(i) Verify Goldbach conjecture for all even n, $2 < n < 100$.

(ii) How many even numbers can you find which have just one decomposition into the sum of two primes?

2.10 Let $S_n = \frac{1}{2} + \frac{1}{3} + \frac{1}{5} + \ldots + \frac{1}{P_n}$.

We have $S_1 = \frac{1}{2}$; $S_2 = \frac{1}{2} + \frac{1}{3} = \frac{5}{6}$; $S_3 = \frac{5}{6} + \frac{1}{5} = \frac{31}{30}$; $S_4 = \frac{31}{30} + \frac{1}{7} = \frac{247}{210}$,

$\ldots$

We notice that at each stage the partial sum S_n consists of an odd integer o over an even integer e. Could this be true in general? Mathematical induction would seem to be called for.

Consider the statement that S_n is always of the form o/e.

$S_1 = 1/2$ is certainly of this form. Assume therefore that S_k is of this form for some k.

Then

$$S_{k+1} = S_k + \frac{1}{P_{k+1}} = \frac{o}{e} + \frac{1}{P_{k+1}} = \frac{oP_{k+1} + e}{eP_{k+1}};$$

now oP_{k+1} is odd and e is even so $oP_{k+1} + e$ is odd and eP_{k+1} is even. So the inductive hypothesis holds for $n = k + 1$ and by induction the result is true for all n.

But a fraction of the form o/e can never be an integer and we are done.

TOPICS FOR INVESTIGATION:

(i) What other rational series can you find whose partial sums can never be an integer?

(ii) How about $\frac{1}{1} + \frac{1}{2} + \frac{1}{3} + \ldots + \frac{1}{n}$?

(iii) How about $\frac{1}{1^2} + \frac{1}{2^2} + \frac{1}{3^2} + \ldots + \frac{1}{n^2}$?
[This has a beautifully lateral solution which you are invited to discover.]

(iv) Just how big can $\frac{1}{2} + \frac{1}{3} + \frac{1}{5} + \ldots + \frac{1}{P_n}$ get as n increases? Can it pass 100 for example?

3 Geometry

3.1

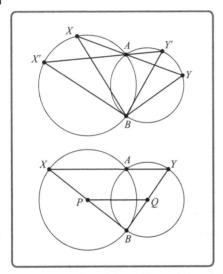

No matter what the position of X and Y, the $\angle YXB$ and $\angle XYB$ are always the same size, because angles on the same chord of a circle are equal. Thus the triangle BXY is always similar to the triangle $BX'Y'$.

So to maximize $|XAY|$, we maximize the area of the triangle BXY and this is achieved by making BX and BY diameters of the two circles. In this case, if P and Q are the centers of the circles, then XAY is parallel to PQ and $|XAY| = 2|PQ|$.

DOI: 10.1201/9781003341468-15

3.2

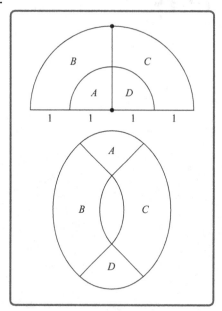

Take the semi-circular piece of glass and cut it into four pieces A, B, C and D as shown in the diagram.

Now rearrange the four pieces as shown to form a nice "elliptical" table with a hole in the middle suitable for an artistic wooden inlay!

TOPIC FOR INVESTIGATION: What other pieces of furniture can you design geometrically?

3.3 Let the equations of the three circles be $S_1 = 0$, $S_2 = 0$ and $S_3 = 0$.

Now what does the equation $S_1 - S_2 = 0$ represent? Since the square terms cancel, the left hand side is a linear expression, and since it clearly passes through the points of intersection of $S_1 = 0$ and $S_2 = 0$, it must be the equation of common chord of the two circles with these equations. Similarly, $S_2 - S_3 = 0$ is the equation of the common chord L_2 of the two circles with equations $S_2 = 0$ and $S_3 = 0$.

Now L_3 has equation $S_1 - S_3 = 0$; but $S_1 - S_3 = (S_1 - S_2) + (S_2 - S_3)$ which is a linear combination, so the lines L_1, L_2 and L_3 are concurrent.

TOPIC FOR INVESTIGATION: What happens if the circles do not intersect?

3.4

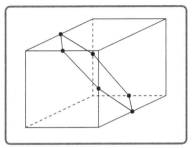

Join the midpoints of the square faces of the cube as shown in the diagram and a perfect regular hexagon results. If you cut the cube along the plane of this hexagon, it is split into the congruent pieces each with a regular hexagonal face.

TOPIC FOR INVESTIGATION: What other solids can you cut to get intersecting plane geometric shapes?

3.5

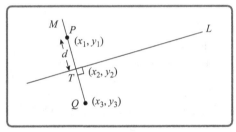

Let L have equation $ax + by + c = 0$ and P have coordinates (x_1, y_1). Any line perpendicular to L has equation $bx - ay + k = 0$, for some constant k. Since P lies on the line M, we have $bx_1 - ay_1 + k = 0$ so $k = ay_1 - bx_1$. Thus M has equation $bx - ay + ay_1 - bx_1 = 0$ and since (x_2, y_2) also lies on M, we have $bx_2 - ay_2 + ay_1 - bx_1 = 0$ or $b(x_2 - x_1) - a(y_2 - y_1) = 0$ (∗). Also, (x_2, y_2) lies on L, so we have $ax_2 + by_2 + c = 0$ which gives

$$a(x_2 - x_1) + b(y_2 - y_1) = -c - ax_1 - by_1,$$
$$\text{so } b(x_2 - x_1) - a(y_2 - y_1) = 0. \quad (∗)$$

Squaring and adding these equations, we get:

$$a^2(x_1 - x_2)^2 + b^2(y_2 - y_1)^2 + 2ab(x_2 - x_1)(y_2 - y_1)$$
$$+ b^2(x_1 - x_2)^2 + a^2(y_2 - y_1)^2 - 2ab(x_2 - x_1)(y_2 - y_1)$$
$$= (ax_1 + by_1 + c)^2$$

So $(a^2 + b^2)[(x_1 - x_2)^2 + (y_1 - y_2)^2] = (ax_1 + by_1 + c)^2$. So

$$d^2 = \frac{(ax_1 + by_1 + c)^2}{a^2 + b^2},$$

which is the usual formula for the distance.

TOPIC FOR INVESTIGATION: Can you use the above technique to find a formula for (x_2, y_2), the coordinates of T, and a formula for (x_3, y_3), the coordinates of Q, the image of P under axial symmetry in L?

3.6

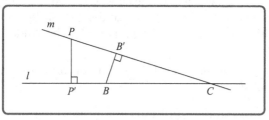

This problem is due to the mathematician James Joseph Sylvester who posed it in 1893. The lateral solution we present is due to Leroy Kelly.

Suppose that a finite set S of points do not all lie on the same straight line. Call a line "connecting" if it contains at least two points of S. Since S is a finite set, there must exist a point P and a connecting line l such that the distance from P to l is greater than zero but is smaller than any other such distance. Let us call a line that contains exactly two points of S an ordinary line. We will show that l is ordinary, by way of contradiction.

Assume that l is not ordinary. Then it contains at least three points of S. At least two of these are on the same side of P', the perpendicular projection of P on l. Call these points B and C, with B being closer to P' (and possibly coinciding with it). Now draw the connecting line m containing P and C, and the perpendicular from B to B' on m. Then $|BB'| < |PP'|$. This follows from the fact that triangles $PP'C$ and $BB'C$ are similar (right angle and common angle C), but $BB'C$ is contained within $PP'C$.

However, this contradicts the original definition of P and l as the point-line pair which are at the smallest positive distance apart and the result is established.

TOPIC FOR INVESTIGATION: Does Sylvester's result hold if the set S is infinite? Can you find a proof or a counterexample?

3.7

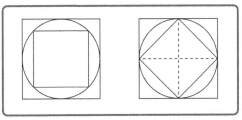

It takes literally a lateral twist or rotation to solve this problem!

Rotating the inner square through a right angle as in the second diagram, we see that the area of the outer square is double the area of the inner square, or 14 square units.

3.8

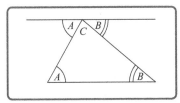

This is one of the most important results in mathematics yet the proof presented in many textbooks is awkward.

For a lateral solution, draw a line through the vertex of the triangle parallel to the base. Since alternate angles are equal, we get at once that $A + B + C = 180°$.

TOPIC FOR INVESTIGATION: Can you find the sum of the internal angles of a quadrilateral? Of an n-sided figure?

3.9

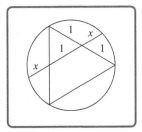

Take an equilateral triangle, of sidelength 2 for convenience, and draw its circumcircle. Take the midpoints of two sides of the triangle, join them and produce this line segment to meet the circumcircle on both sides. By symmetry, the two segments marked x have the same length. Now by the intersecting chords theorem, we have $x(1 + x) = 1 \cdot 1 = 1$, so $x + x^2 = 1$ or $x^2 + x - 1 = 0$.

By the quadratic formula,

$$x = \frac{-1 \pm \sqrt{1^2 - 4(1)(-1)}}{2} = \frac{-1 \pm \sqrt{5}}{2}.$$

Since x is a positive distance, we must take $x = \frac{-1 + \sqrt{5}}{2}$, so

$$x + 1 = \frac{2 - 1 + \sqrt{5}}{2} = \frac{\sqrt{5} + 1}{2} = \phi.$$

TOPIC FOR INVESTIGATION: Show how to construct ϕ using a regular pentagon. What other geometric constructions for ϕ can you find?

3.10

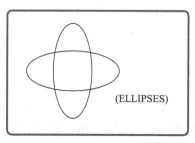

(ELLIPSES)

A closed orbit of a satellite under gravitational attraction is always an ellipse.

Four observations are not enough because two completely different ellipses can pass through the same four points, as in the diagram.

However, if a fifth observation is taken, this will determine the precise orbit of the object.

TOPIC FOR INVESTIGATION: Who first discovered that the orbit of a body under an inverse square law of attraction is a conic section? What techniques did he use to achieve this fantastic result?

4 Trigonometry

4.1 We rewrite the expression as

$$\left(\frac{\tan 45° + \tan 15°}{1 - \tan 45° \tan 15°}\right)^2 = E,$$

exploiting the fact that $\tan 45° = 1$ – this is a lovely lateral twist. Now using the fact that

$$\tan(A + B) = \frac{\tan A + \tan B}{1 - \tan A \tan B},$$

we have $E = [\tan(45° + 15°)]^2 = [\tan 60°]^2 = (\sqrt{3})^2 = 3$.

Topic for Investigation: What other clever ways can you write 1?

4.2 We have $\tan(180° - X) = -\tan X$, if X is not a right angle. Since $A + B + C = 180°$, we have $\tan(A + B) = \tan(180° - C) = -\tan C$, so

$$\frac{\tan A + \tan B}{1 - \tan A \tan B} = -\tan C,$$

which simplifies to $\tan A + \tan B + \tan C = \tan A \tan B \tan C$. So we seek three positive integers a, b and c such that $a + b + c = abc$.
Assume that $0 < a \leq b \leq c$ for a, b, c in $\mathbb{N}$. Then $abc = a + b + c \leq 3c$, so $ab \leq 3$, and $(a, b) = (1, 1)$, $(1, 3)$ or $(1, 2)$.

(i) $(a, b) = (1, 1)$ gives $1 \cdot 1 \cdot c = 1 + 1 + c$ or $c = c + 2$. Not possible.

(ii) $(a, b) = (1, 3)$ gives $1 \cdot 3 \cdot c = 1 + 3 + c$ or $3c = c + 4$ or $c = 2$. Then $b = 3$, $c = 2$, and $b > c$. Not possible.

(iii) $(a, b) = (1, 2)$; $1 \cdot 2 \cdot c = 1 + 2 + c$, so $c = 3$ and this fits. So $(a, b, c) = (1, 2, 3)$ and this is the unique solution.

Topic for Investigation: What other trigonometric identities can you connect with number theory? For what angles are $\sec^2 \theta$ and $\csc^2 \theta$ both integers?

DOI: 10.1201/9781003341468-16

4.3

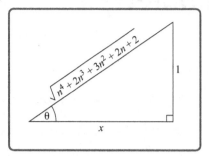

This really looks a mess – and it is. A good lateral principle is, when you have a mess, it is worthwhile spending some time tidying it up.

The algebraic term under the square root sign is in sore need of simplifica-tion – can we transform it into something else? A moment's thought shows that an inverse cosine is equally hopeless. How about an inverse tangent? Draw a right-angled triangle as in the diagram. Then $x^2 + 1^2 = n^4 + 2n^3 + 3n^2 + 2n + 2$ or $x^2 = n^4 + 2n^3 + 3n^2 + 2n + 1$, from which $x = n^2 + n + 1$, very conveniently, so

$$\theta = \tan^{-1}\left(\frac{1}{n^2 + n + 1}\right).$$

Now a lovely lateral twist – this can be written as

$$\tan^{-1}\left(\frac{(n + 1) - n}{1 + n(n + 1)}\right) = \tan^{-1}(n + 1) - \tan^{-1} n.$$

So our finite sum can be written as a telescoping sum:

$$\tan^{-1}(n + 1) - \tan^{-1}(n)$$
$$\tan^{-1}(n) \quad -\tan^{-1}(n - 1)$$
$$\tan^{-1}(n - 1) - \tan^{-1}(n - 2)$$
$$\vdots \qquad \vdots$$
$$\tan^{-1}(4) \quad -\tan^{-1}(3)$$
$$\tan^{-1}(3) \quad -\tan^{-1}(2)$$
$$\tan^{-1}(2) \quad -\tan^{-1}(1)$$

Summing and canceling, we get $\tan^{-1}(n+1) - \tan^{-1}(1) = \tan^{-1}(n+1) - \pi/4$. Letting $n \to \infty$, we get $(\pi/2) - (\pi/4) = \pi/4$.

TOPIC FOR INVESTIGATION: What other telescoping sums can you use to find sums of trigonometric series?

4.4 It is clear that $\tan A = 1/3$, $\tan B = (1/2)$ and $\tan C = 1$.

So $A + B = \tan^{-1}(1/3) + \tan^{-1}(1/2) = \tan^{-1}\left(\frac{(1/3)+(1/2)}{1-(1/3)(1/2)}\right)$
$= \tan^{-1}((5/6)(5/6)) = \tan^{-1}(1) = C = \pi/4$. So $A + B = C$.

<u>TOPIC FOR INVESTIGATION</u>: Show how this very simple situation can be used to calculate the value of π.

What can you say about $\tan^{-1}(1/1) + \tan^{-1}(1/2) + \tan^{-1}(1/4) + \dots + \tan^{-1}(1/n)$?

Does $\sum_{n=1}^{\infty} \tan^{-1}(1/n)$ converge?

4.5 We use the well-known cosine formula for a triangle. This gives

$$\cos A = \frac{b^2 + c^2 - a^2}{2bc} = \frac{25 + 16 - 36}{2 \cdot 5 \cdot 4} = \frac{5}{40} = \frac{1}{8}.$$

Next,

$$\cos C = \frac{5^2 + 6^2 - 4^2}{2 \cdot 5 \cdot 6} = \frac{45}{60} = \frac{3}{4}.$$

Now $\cos 2C = 2\cos^2 C - 1 = \frac{2 \cdot 9}{16} - 1 = \frac{1}{8}$. Thus $\cos A = \cos 2C$, so $A = 2C$.

<u>TOPICS FOR INVESTIGATION</u>: Properties of the $(3, 4, 5)$ triangle are well known; can you find a property of the $(2, 3, 4)$ triangle? The $(1, 2, 3)$ triangle?

Can you count all the "different" kinds of integer-sided triangles with perimeter 12?

4.6

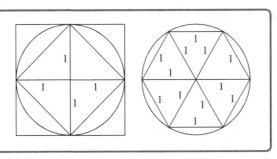

Take a unit circle and consider its inscribed and circumscribed squares. We have that the outer square area is greater than the circle area which is in turn greater than the inner square area.
Thus $4 > \pi \cdot 1 > 2$, so $4 > \pi > 2$.
How about perimeters?
We have $8 > 2\pi r > 4\sqrt{2}$ and $r = 1$, so $4 > \pi > 2\sqrt{2} = 2.818\dots$
Good, but not there yet!
Now consider a regular hexagon inscribed in a unit circle. We have $2\pi r > 6 \cdot 1 = 6$, so $\pi > 3$ and we have shown $4 > \pi > 3$.

TOPICS FOR INVESTIGATION: Look at the incredible work of Archimedes, the greatest mathematician of antiquity (and some would say of all time) who used a 96-sided regular polygon to show that $3\frac{1}{7} > \pi > 3\frac{10}{71}$, a truly remarkable achievement.

Bounding areas inside and outside is certainly valid, but does this technique always work for perimeters?

4.7

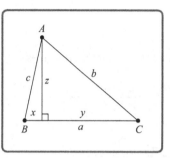

Let z be the length of the perpendicular from A onto BC dividing the side BC into two lengths x and y.

Then $\cot B = x/z$ and $\cot C = y/z$.

So $\cot B + \cot C = (x + y)/z = a/z$ which gives $z = a/(\cot B + \cot C)$.

Now the area of the triangle $\Delta = (1/2)az = (1/2)\dfrac{a^2}{\cot B + \cot C}$.

TOPIC FOR INVESTIGATION: Think of an important problem in surveying where this formula is very useful.

Show that each of the other congruence relations for triangles leads to a formula for the area of a triangle.

4.8

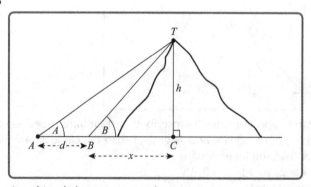

Our friend the cotangent function comes to our rescue again. Let C be the foot of the perpendicular from T onto AB and let $|BC| = x$. If h is the height of the mountain, then $x/h = \cot B$ and $(d + x)/h = \cot A$. So $\cot A - \cot B = (d + x)/h - x/h = d/h$, and $h = d/(\cot A - \cot B)$.

TOPICS FOR INVESTIGATION: Notice we do not know the value of x and need not: it disappears in the wash.

Find other applications of trigonometry in surveying and lateral ways of deriving them.

4.9 This one is very lateral and very pretty, but don't take it too seriously!

If $e^{i\theta} = \cos\theta + i\sin\theta$, then $e^{-i\theta} = \cos\theta - i\sin\theta$, so $1 = e^0 = e^{i\theta-i\theta} = (\cos\theta + i\sin\theta)(\cos\theta - i\sin\theta) = \cos^2\theta + \sin^2\theta$.

TOPIC FOR INVESTIGATION: Prove as many trigonometric identities as you can using Euler's identity. How valid are these "proofs"?

4.10

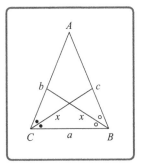

First we need a formula for the length of an angle bisector x. We have

$$(1/2)xb \sin(C/2) + (1/2)xa \sin(C/2) = (1/2)ab \sin C.$$

This gives

$$x = \frac{2ab}{a+b} \cos(C/2).$$

So we have

$$\frac{2ab}{a+b} \cos(C/2) = \frac{2ac}{a+c} \cos(B/2),$$

or

$$(a+c)b \cos(C/2) = (a+b)c \cos(B/2).$$

Using well-known formulae, and putting $2s = a + b + c$, we get

$$(a+c)b \sqrt{\frac{s(s-c)}{ab}} = (a+b)c \sqrt{\frac{s(s-b)}{ac}}$$

which quickly reduces to

$$b(a+c)^2(a+b-c) = c(a+b)^2(a+c-b)$$

After a little healthy calculation, this becomes

$$(b-c)[a^3 + a^2(b+c) + 3abc + bc(b+c)] = 0.$$

As the second term is clearly positive, we have $(b - c) = 0$, so $b = c$ and the triangle is isosceles.

TOPICS FOR DISCUSSION: This is the Steiner-Lehmus theorem, notoriously tricky to prove. Read the literature about it, especially the search for a "direct" proof.

Can you say anything about triangles with two equal <u>external</u> angle bisectors? What about equal medians and altitudes?

5 Probability

5.1 A good lateral principle is "think a bit before you plunge into calculations"! Take an orange for example and mark any three distinct points on its surface. What do you notice? It is easy to cut the orange in two halves so that all three points lie on the same hemisphere.

This is true for all spheres – take the great circle through any two of the points and slice the sphere along this line, passing through the center of the sphere. The third point now lies on one of these two hemispheres (or both!). So the probability is 1.

TOPIC FOR INVESTIGATION: What happens if we take four distinct points?

5.2 The probability of choosing an odd number is 1/2. For the rectangle to have odd area, its length and its width must be odd. So the probability of the area of the rectangle being odd is $(1/2)(1/2) = (1/4)$.

TOPIC FOR INVESTIGATION: Write down the dimensions of rectangles of small size, that is, 1×1, 1×2, 1×3, 2×2, 2×3, etc. and see how the theory and practice agree.

5.3 *A*, being a lady or a gentleman, cannily allows *B* to nominate first. The strange thing is that whatever triple *B* nominates, *A* can nominate a triple with a higher probability of winning! Here are the details:

B chooses	A chooses	Odds in favor of A
HHH	THH	7 to 1
HHT	THH	3 to 1
HTH	HHT	2 to 1
HTT	HHT	2 to 1
THH	TTH	2 to 1
THT	THH	2 to 1
TTH	HTT	3 to 1
TTT	HTT	7 to 1

The rule is that the second player starts with the opposite of the middle choice of the first player and then follows with the first player's first two choices.

DOI: 10.1201/9781003341468-17

57

The "game" is known as Penney's game. Don't play it unless you are A!

Topic for Investigation: Can you formulate similar strategies for two tosses of a coin or four tosses?

Discuss the ethics of playing a game where the odds are in your favor and your opponent is not aware of this fact.

5.4

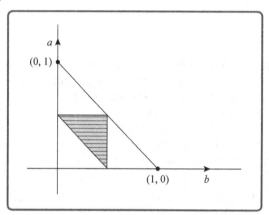

The triangle inequality states that if a, b and c are the lengths of the sides of a triangle, then $a + b > c$, $b + c > a$ and $a + c > b$, and that if any triple of positive real numbers (a, b, c) fails to satisfy all three of these inequalities, then there is no triangle with sidelengths a, b and c. Of course if the triple (a, b, c) does satisfy all these inequalities, then we can easily construct (just one) triangle with sidelengths a, b and c.

Suppose the rod has unit length and that the three pieces have lengths a, b and $1 - a - b$. Then the space of all possible values of (a, b) in the $a - b$ plane is the triangle formed by the intersection of the half planes $a > 0$, $b > 0$ and $a + b < 1$. We assume that each coordinate pair in this space is equally likely and that the probability that a triangle can be formed is the ratio of the area containing all pairs (a, b) which a triangle can be formed, to the area of the entire triangle, which is $1/2$. Now apply the triangle inequality to a, b and $1 - a - b$ to get $a + b > 1 - a - b$, $b + 1 - a - b > b$ and $b + 1 - a - b > a$ which simplify to $a + b > 1/2$, $b < 1/2$ and $a < 1/2$. The intersection of these three half-planes gives a triangle of area $1/8$ (shaded) so the required probability is $(1/8)(1/2) = 1/4$.

Topics for Investigation: Try constructing a triangle with sidelengths $(1, 2, 4)$. Can you do it?

How do you construct a triangle with sidelengths $(2, 3, 4)$? How many different triangles do you get?

Is there a quadrilateral inequality?

If a thin rod is broken in three places into four pieces, what is the probability that the four pieces can form a quadrilateral?

5.5

A\B	2	2	4	4	9	9
1	A	A	A	A	A	A
1	A	A	A	A	A	A
6	B	B	B	B	A	A
6	B	B	B	B	A	A
8	B	B	B	B	A	A
8	B	B	B	B	A	A

B\C	1	1	6	6	8	8
3	C	C	B	B	B	B
3	C	C	B	B	B	B
5	C	C	B	B	B	B
5	C	C	B	B	B	B
7	C	C	C	C	B	B
7	C	C	C	C	B	B

C\A	3	3	5	5	7	7
2	C	C	C	C	C	C
2	C	C	C	C	C	C
4	A	A	C	C	C	C
4	A	A	C	C	C	C
9	A	A	A	A	A	A
9	A	A	A	A	A	A

Playing A against B, as in diagram 5(i), A wins 20 times while B wins 16 times. So the probability of A winning is 20/36 = 5/9, and the probability of B winning is 16/36 = 4/9.

Playing B against C as in diagram 5(ii), B wins 20 times while C wins 16 times. So the probability of B winning is 20/36 = 5/9, while the probability of C winning is 16/36 = 4/9.

Finally, playing C against A, as in diagram 5(iii), C wins 20 times while A wins 16 times. So the probability of C winning is 20/30 = 5/9, while the probability of A winning is 16/36 = 4/9.

So we have this extraordinary situation – A beats B consistently, B beats C consistently, but C beats A consistently! For numbers, if $x > y$ and $y > z$, then we have $x > z$, which is called transitivity. The situation we have here is called intransitive dice. Most people do not believe this astonishing paradox at first sight and keep checking the boxes to see if there has been a miscount, but the numbers do not lie; probability does not behave in the way that one expects.

Of course in real life it often happens that football team A beats team B, and B beats C, while C turns round and beats A, but we do not expect this type of thing in mathematics! There is also the popular game of paper, rock and scissors, where paper covers rock, rock smashes scissors and scissors cuts paper.

The real danger is that you are invited to play a game with a dice sharp, who gallantly allows you to choose A, B or C. Whichever die you choose, he can choose a die to beat you with – choose A, he chooses C; choose B, he chooses A; choose C, he chooses B. So beware!

TOPIC FOR INVESTIGATION: What happens if three people roll A, B and C simultaneously (the highest score to be the winner)? Which is the best die to have now, or are they all equal?

5.6 The possible outcomes are HHH, <u>HHT</u>, <u>HTH</u>, <u>THH</u>, HTT, THT, TTH, TTT, where the favorable outcomes are underlined. Probability is number of favorable outcomes divided by total number of outcomes $= 3/8$.

The more lateral approach however is to look elsewhere.
We have $(H + T)^3 = H^3 + 3H^2T + 3HT^2 + T^3$, either by the binomial theorem or ordinary algebra. The H^2T term represents two heads and one tail and the probability is $3/(1 + 3 + 3 + 1) = 3/8$.

TOPIC FOR INVESTIGATION: What is the probability of having at least two tails?

In practice, why is the probability of having a tail very slightly greater than having a head?

5.7 Suppose your father was born in January and your mother was born in November. Now January to November is more than six months, but birthdays come in a cycle and November to January is less than six months. So in general the probability is in fact 1, or certainty!

TOPIC FOR INVESTIGATION: In a game of Bridge or Whist, which probability is greater – that a given pair will, between them, hold all the trumps, or none of the trumps?

5.8

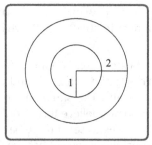

If the point lies within the inner disc, it is nearer the center than the circumference. The desired probability is therefore the area of the inner disc divided by the area of the whole disc.

This is $\pi(1)^2/\pi(2)^2 = 1/4$.

TOPIC FOR INVESTIGATION: How could one refute the solution that since the point can lie anywhere along a radius of length 2, the probability is 1/2?

5.9

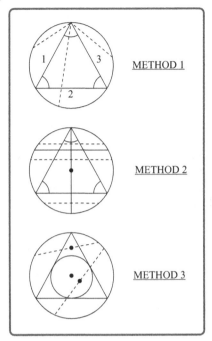

METHOD 1

METHOD 2

METHOD 3

This really is a very tricky problem that some people almost come to blows over! The answer all depends on what one means by "random" and many different answers are possible depending on how a "random" chord is generated. This problem is so significant that it has been given a specific name – Bertrand's paradox. The lateral dimension is the number of ways in which the random chord can be generated.

Method 1 Choose one vertex of the inscribed equilateral triangle as one endpoint of the chord. If the other endpoint lies in segment 1 or segment 3, the length of the chord is less than the side of the equilateral triangle, while if it lies in segment 2, the length of the chord is greater than the side of the equilateral triangle. But the curved perimeters of the three segments all have equal length, so the desired probability is 1/3.

Method 2 Draw a diameter of the circle which is perpendicular to a side of the inscribed equilateral triangle. Now draw (dotted) chords perpendicular to this diameter. By this method it emerges that the probability that a random chord is longer than the side of the equilateral triangle is 1/2.

Method 3 Pick any point within the disc and construct a chord with that point as its midpoint. The length of the chord is longer than the side of the inscribed equilateral triangle if and only if the chosen point lies inside a concentric disc whose radius is half of the original disc. The required probability, therefore the ratio of the areas of these two discs or 1/4.

TOPICS FOR DISCUSSION: Thinking laterally, can you find another approach to randomness of a chord which gives yet another different probability?

You are given three distinct points A, B and C in the plane which do not lie on a straight line. What is the probability that the triangle ABC is acute-angled? Is obtuse-angled? Is right-angled?

5.10 One of the strongest weapons in a lateral thinker's armoury is the thought that you may have solved a problem before, perhaps in a different guise. Look back to problem 5.6 about coin tossing. There the probability of heads or tails is the same – (1/2) in each case. Now the probability of a child being a boy or a girl is also (1/2) (very closely!), so this time we look at $(B + G)^4 = B^4 + 4B^3G + 6B^2G^2 + 4BG^3 + G^4$. The total number of possibilities is $1 + 4 + 6 + 4 + 1 = 16$ and the favorable possibilities are given by the $4BG^3$ term which has coefficient 4.

So the probability that in a family of four children there are three girls and one boy is $4/16 = 1/4$. It's a toss-up!

TOPICS FOR INVESTIGATION: Consider how binary numbers with, say, six digits, for example 101101. What is the probability that such a number has two ones and four zeroes?

Can you find yet another scenario in which the same principles apply?

6 Combinatorics

6.1

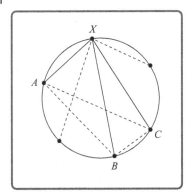

The lateral twist one can use here is to translate the problem to geometry, called graph theory in this context.

Imagine the people represented by six points on the circumference of a circle. If two people know each other, join their dots by a full line segment. If they are strangers to each other, join their dots by a dotted line.

Now consider one of the points X, which is joined to five other points. By the pigeonhole principle, at least three of these lines must be of the same kind – say full lines for convenience (if there are three dotted lines, the reasoning is similar). Say the full lines are XA, XB and XC.

If any of the lines AB, AC or CB is a full line, we have a triangle with all its sides full lines, so we have three people each of whom knows the other two. Otherwise ABC is a triangle with dotted line sides, giving us three people neither of whom knows either of the other two.

This is regarded as one of the prettiest lateral proofs in mathematics.

TOPIC FOR INVESTIGATION: What happens for $2, 3, 4, 5, \ldots, n$ people? Find out about Ramsey theory, a rather difficult branch of mathematics which examines problems like this.

6.2 One has to be careful to seek a method of solution which does not get out of hand numerically. Divide and conquer is a standard lateral technique – the trick is to split the problem into subcases, each of which can be solved neatly and individually and hopefully in a similar way.

DOI: 10.1201/9781003341468-18

In this version, we include zero as a digit but not as an initial digit. Our subcases are determined by the number of digits in the number.

One digit: There are clearly 9 of these.

Two digits: We have 9 choices for the first digit (because the number cannot begin with 0) but we again have nine choices for the second digit, so we have $9 \cdot 9 = 81$ choices.

Three digits: We have $9 \cdot 9 \cdot 8 = 648$ choices.

Four digits: We have $9 \cdot 9 \cdot 8 \cdot 7 = 4,536$ choices.

Five digits: We have $9 \cdot 9 \cdot 8 \cdot 7 \cdot 6 = 27,216$ choices.

Six digits: We have $9 \cdot 9 \cdot 8 \cdot 7 \cdot 6 \cdot 5 = 136,080$ choices.

Seven digits: We have $9 \cdot 9 \cdot 8 \cdot 7 \cdot 6 \cdot 5 \cdot 4 = 544,320$ choices.

Eight digits: We have $9 \cdot 9 \cdot 8 \cdot 7 \cdot 6 \cdot 5 \cdot 4 \cdot 3 = 1,632,960$ choices.

Nine digits: We have $9 \cdot 9 \cdot 8 \cdot 7 \cdot 6 \cdot 5 \cdot 4 \cdot 3 \cdot 2 = 3,265,920$ choices.

Ten digits: We have $9 \cdot 9 \cdot 8 \cdot 7 \cdot 6 \cdot 5 \cdot 4 \cdot 3 \cdot 2 \cdot 1 = 3,265,920$ choices.

This gives a total of 8,877,690 choices (please check carefully – your author has been known to make numerical mistakes!)

TOPICS FOR INVESTIGATION: What is the answer if we exclude zero altogether? What can you relate this to?

How about finding the sum of those 8,877,690 numbers above?

6.3 The worst possible scenario is that the first three socks you withdraw from the drawer are red, blue and green in some order. But the fourth sock must be one of these colors and now you have a matching pair. So you must withdraw at most four socks (and at least two) to ensure a matching pair. If you want a pair of blue socks, if you are unfortunate, the first ten socks you withdraw could all be red or green. In this situation, you need two extra withdrawals to ensure a blue pair. So the answer this time is at most 12 socks (and at least 2).

TOPICS FOR INVESTIGATION: Can you generalize the problems for x red socks, y blue socks and z green socks?

What about somebody, like your author, who is red-green color blind? How does that change things?

Suppose some of the socks have a hole in them and you are prepared to wear a same color pair with at most one holey sock?

Mathematics is almost infinitely variable and inventive, and you can make up your own conditions!

My wife Anne, by the way, thinks that the most lateral solution is to switch the light on or fetch a torch!

6.4

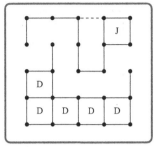

If David joints the dotted line, Jane does not have to take any boxes – she can join the new dotted line. Any move David now makes will result in Jane's winning all the remaining boxes, except possibly two, and thus will win the game.

But be careful – your author once played this move in a game against his brother and nearly got beaten up!

TOPIC FOR INVESTIGATION: Think carefully about other such games such as noughts and crosses or chequers and see if you can come up with clever lateral moves to turn the tables.

6.5 The lateral trick here is to use the equation $2^{a+b} = 2^a \cdot 2^b$ (the basis of logarithms) which turns addition into multiplication. So turn the given table into powers of 2 as follows:

2^4	2^9	2^2
2^3	2^5	2^7
2^8	2^1	2^6

or

16	512	4
8	32	128
256	2	64

which automatically fulfils our requirements. Using other bases such as three gives us infinitely many examples, for example:

81	19683	9
27	243	2187
6561	3	729

TOPICS FOR INVESTIGATION: Can you exploit the fact that $2^0 = 1$ to find an example with smaller numbers?

And surely you will want to find the example with the smallest possible numbers, not necessarily all powers of 2.

6.6 The books of A and B must never be together, so the lateral twist is to tie them together and treat AB as a single object! With the ten other objects, this gives 11! different permutations; but we could also have bound them together as BA, which gives 11! additional permutations. So the total number of permutations where A and B are not adjacent is the overall total number of permutations of the 12 objects (12!), minus the number where A and B are adjacent, which is 2(11!). Hence the number required is 12! – 2(11!). You are invited to work this number out by hand – this activity is good for your brain.

TOPICS FOR INVESTIGATION: Suppose we again have twelve books and three particular books A, B and C. How many permutations are possible if no two of A, B and C are allowed to be adjacent? If at least two of A, B and C must not be adjacent? If all three of A, B and C cannot be adjacent?

Suppose that of two particular books X and Y, X must always be to the left of Y, but not necessarily immediately to the left of Y. How many different arrangements are now possible?
[This one has a beautifully lateral solution – think probability!]

6.7 Each snap creates exactly one extra piece, therefore to break a bar with n squares into its component small square pieces will require $n - 1$ snaps. So here the answer is 34 snaps. A classic lateral solution.

TOPIC FOR INVESTIGATION: What connection can you find between tennis matches, jigsaw puzzles and chocolate bars?

6.8 In many problems it is instinctive to step down in size. Consider a 2×2 antimagic square, consisting of the numbers 1, 2, 3 and 4. Adding distinct pairs from $\{1, 2, 3, 4\}$ one gets $\{3, 4, 5, 6, 7\}$ or five distinct sums. But an antimagic 2×2 square requires six distinct sums, so the problem has no solution. This is a very beautiful lateral solution – a true classic.

9	8	7
2	1	6
3	4	5

Here is a 3×3 antimagic square using the numbers $\{1, 2, 3, 4, 5, 6, 7, 8, 9\}$. Not the spiral configuration, starting at 1 in the center, and the positioning of the odd and even numbers. This looks easy but is quite tricky.

1	2	3	4
5	6	7	8
9	10	11	12
13	14	16	15

Next, we look at the 4 × 4 case using the numbers
{1, 2, 3, 4, 5, 6, 7, 8, 9, 10, 11, 12, 13, 14, 15, 16}. This is of course more difficult, but more interesting. How about arranging the numbers in numerical order, starting from the left in each row? This <u>almost</u> works, but both diagonals must sum to 34. The lateral twist is to swap 15 and 16 and the new configuration miraculously works.

3	5	11	10
16	6	8	4
13	12	9	2
1	15	7	14

Sometimes in mathematics one can add extra conditions to a problem and still solve it. Laterally this can be done by observing all the solutions and seeing that they satisfy additional conditions. Suppose we ask for a 4 × 4 antimagic square where the ten totals are consecutive integers, again using only the integers 1 to 16? Amazingly, this can still be done and in several ways. The totals involved are either 29 to 38 or 30 to 39.

3	5	11	10
16	6	8	4
13	12	9	2
1	15	7	14

Topics for Investigation: One could spend a lifetime on this fascinating topic, so we must restrain ourselves!

(i) Can you find a 3 × 3 antimagic square based on the numbers 1 to 9 where all 10 rows, columns and diagonals sum to different totals which are consecutive integers?

(ii) Can the "spiral technique" be used to generate a 4 × 4 magic square?

(iii) Do solutions exist for $n > 4$?

(iv) Does a "doubly antimagic" square exist which works for both addition and multiplication?

(v) Note that the famous 3 × 3 <u>magic</u> square is multiplicatively antimagic – a nice bonus.

4	9	2
3	5	7
8	1	6

6.9 This, to my mind, is one of the most important facts in mathematics, central to a whole lot of theory and applications. In our lateral approach we give three very different proofs.

PROOF 1 Given a subset T of a set S, an element x of S either lies in T or it does not lie in T and one and only one of these possibilities holds. So if we wish to form a subset T of S, for each x in S there are precisely two choices – either x lies in T or it does not (take it or leave it!). Every time we make a different choice, we get a different subset T of S. Since S has exactly n elements and there are exactly two choices for each element, by elementary combinatorics, there are $2 \cdot 2 \cdot 2 \cdot \ldots \cdot 2 = 2^n$ choices we can make and so there are 2^n distinct subsets T of S.

[When we choose each element x of S in T, we get the whole set S, and when we choose no element x of T we get the empty set $\varnothing$.]

PROOF 2 In this proof we concentrate on the size or cardinality of the set T, so let $|S| = n$ and then $|T| = 0, 1, 2, \ldots, n$. Now there are $^nC_0$ (spoken as "n choose zero") ways of choosing a subset T with 0 elements (the empty set); $^nC_1$ ways of choosing a subset T with a single element;..., ^nC_r ways of choosing a subset T with exactly r elements, and finally ^nC_n ways of choosing a subset T with exactly n elements ($= S$).

Moreover, these choices all independent of each other and cover all cases, so the total number of ways, and also the total number of subsets, is given by

$$^nC_0 + {}^nC_1 + {}^nC_2 + \ldots + {}^nC_r + \ldots + {}^nC_n$$

Now evaluating this sum still looks difficult so let's laterally think in another box. Where have we seen those numbers before? They are the binomial coefficients, and come to think of it, choosing subsets is a bit like computing powers and adding the results. We have

$$(1 + x)^n = {}^nC_0x^0 + {}^nC_1x^1 + \ldots + {}^nC_rx^r + \ldots + {}^nC_nx^n$$

$$= \sum_{r=0}^{n} {}^nC_rx^r.$$

Putting $x = 1$ on both sides, we get $2^n = \sum_{r=0}^{n} {}^nC_r$, as required.

PROOF 3 Sadly, the previous two proofs, while very pretty, are not completely watertight. Proof 1 contains the infamous $2 \cdot 2 \cdot 2 \cdot \ldots \cdot 2$ which, while widely used in mathematics, is not properly defined. And in Proof 2, how is one to prove the binomial theorem?

Lurking behind Proofs 1 and 2 is one of the most useful, ingenious, and indeed lateral of all mathematical techniques – mathematical induction. In the case of proof by induction, it runs as follows:

Let $P(n)$ be a proposition for each natural number n.

Suppose that $P(1)$ is a true proposition. Suppose also that k is a natural number and that whenever $P(k)$ is a true proposition, then $P(k + 1)$ is a true proposition also.

Then $P(n)$ is a true proposition for all n in $\mathbb{N}$.

This method of proof by induction seems very plausible but of course it is an axiom, assumed without proof.

So, after all that, how about a proper proof by mathematical induction that a set S with exactly n elements has precisely 2^n distinct subsets?

Let $P(n)$ be the proposition that for each n in $\mathbb{N}$, every set S with $|S| = n$ has precisely 2^n distinct subsets.

Now $P(1)$ is the proposition that a set S with exactly one element x, $S = \{x\}$ has precisely $2^1 = 2$ distinct subsets. These subsets are $\{x\}$ and the empty set $\varnothing$. So $P(1)$ is a true proposition. Suppose now that $P(k)$ is true for some positive integer k – that is, any set with cardinality k has precisely 2^k distinct subsets.

Consider now a set S of cardinality $k + 1$. We isolate a definite element x of S. The subsets of S are of two kinds – those which do not contain x and those which do contain x. The set $S - \{x\}$ has cardinality k and so by our inductive assumption has precisely 2^k distinct subsets. To form a subset of S containing x, we adjoin $\{x\}$ to any of the subsets of $S - \{x\}$. There are precisely 2^k of these subsets also. Thus we have $2^k + 2^k = 2 \cdot 2^k = 2^{k+1}$ distinct subsets of S, so our proposition $P(k + 1)$ is true. By the principle of mathematical induction, $P(n)$ is true for all natural numbers n.

TOPICS FOR INVESTIGATION:

Is a subset of a set with n elements more likely to have an odd or an even number of elements?

Can you relate all of this to binary numbers?

Investigate the famous Pascal triangle.

```
                    1
                1       1
            1       2       1
        1       3       3       1
    1       4       6       4       1
    ...     ...     ...     ...     ...
```

6.10 Let us go on a small voyage of discovery and examine naturally a little theory that will help us solve this problem (and many others). Step 1 is, invariably, some data. Let $d(n)$ denote the number of distinct divisors of a positive integer n, including 1 and n itself. By experiment and calculation we find

n	1	2	3	4	5	6	7	8	9	10	11	12	13	14	15	16
$d(n)$	1	2	2	3	2	4	2	4	3	4	2	6	2	4	4	5

We quickly see that if p is a prime number, then $d(p) = 2$. [If $d(n) = 2$, must n be prime? You decide.] Also, $d(p^2) = 3; d(p^3) = 4; d(p^4) = 5, \ldots$ and it is not hard to see that $d(p^a) = a + 1$, where p is a prime.

A useful lateral technique is to break a problem into simpler parts; remember that every positive integer n can be expressed as a product of prime power factors, essentially in just one way. So we write

$$n = p_1^{a_1} p_2^{a_2} \ldots p_r^{a_r},$$

where the p_i are distinct prime numbers.

Now, looking at $d(6)$, $d(10)$ and $d(12)$ in the aforementioned table above, one might guess that $d(ab) = d(a)d(b)$ where a and b have no factor in common except 1. This turns out to be the case and is not difficult to prove. So, in general, the total number of divisors $d(n)$ is given by $(a_1 + 1)(a_2 + 1)(\ldots)(a_r + 1)$.

Now here is the final lateral step. We express the given number $1^1 2^2 3^3 4^4 5^5 6^6$ in prime power form.

It becomes $2^{16} \cdot 3^9 \cdot 5^5$ and

$$d(n) = (16 + 1)(9 + 1)(5 + 1)$$
$$= 1020.$$

Notice we can ignore 1^1, which does not contribute to the count.

TOPICS FOR INVESTIGATION: Can you find the sum of all these divisors? In general?

Need $d(ab)$ be equal to $d(a)d(b)$ always? Can you find a proof or a counterexample?

In the table for $d(n)$, does each number appear infinitely often?

7 Dissections

7.1

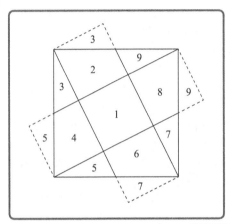

Nine pieces – too many squares.

Four squares – too few pieces.

So we need something in between. Lateral means sideways, so let's shift to the side. After some experimentation, we come up with the following scenario: join the midpoints of the sides of the squares to the opposite corners. 1 is already a square; $2 \cup 3$ form a square; $4 \cup 5$ form a square; $6 \cup 7$ form a square; $8 \cup 9$ form a square.

And so we have our five equal squares.

TOPIC FOR INVESTIGATION: Can we generalize? We want to dissect a square into $n(\neq m^2)$ pieces which can be reassembled to form m squares of equal size. What happens if we look at the trisection points of the sides and join them appropriately?

DOI: 10.1201/9781003341468-19

7.2

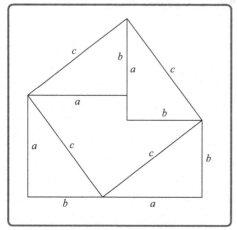

This diagram, which is perfectly general when $a \neq b$ and $a^2 + b^2 = c^2$, shows how to solve the problem.

TOPIC FOR INVESTIGATION: Some people regard this as a dissection "proof" of the theorem of Pythagoras. Would you agree?

What happens if $a = b$? Explore the topic of "degeneracy" in mathematics.

7.3

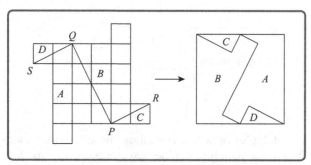

This is a beautiful problem with a beautiful solution, but it requires some lateral thinking to solve it!

The original diagram has 20 small squares, so if these are to be put together to form a large square, its side must be of length $\sqrt{20}$ units.

In the diagram, $2^2 + 4^2 = 20 = |PQ|^2$, so PQ would be a suitable side for the square. Also SQ is perpendicular to PQ as is RP, providing us with the right angles we need. The pieces A, B, C and D now fit together beautifully and symmetrically to form the square we need. Lateral magic!

TOPIC FOR INVESTIGATION: Do you think you can generalize this problem for larger figures similar to the given one? Experiment, experiment, experiment!

7.4

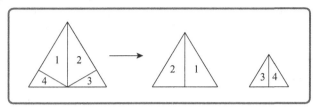

Here is a very simple way of solving the problem.

TOPICS FOR INVESTIGATION: There is another very beautiful solution based on $3^2 + 4^2 = 5^2$. Here are some clues to help you find it:

Take an equilateral triangle of sidelength 5 units and divide it into 25 congruent equilateral triangles. The top 9 triangles form an equilateral triangle of side 3 units.
Now see if you can dissect the remaining piece into three pieces and reassemble them to form an equilateral triangle of side four units.

Do you think it might be possible to dissect an equilateral triangle into just <u>four</u> pieces and reassemble those four pieces into <u>three</u> equilateral triangles, no two of the same size? Surely not! But the answer may surprise you!
[Consider $7^2 = 2^2 + 3^2 + 6^2$.]

7.5 We use the method of proof by contradiction – imagine that we have dissected a solid cube into a finite number $n(\geq 2)$ of solid cubes, no two of the same size. If this assumption leads to a contradiction, we are forced to conclude that our original assumption is false and that no solution to the problem is possible. As the great G.H. Hardy has said, "a chess player will sometimes sacrifice a single piece; the mathematician is prepared to sacrifice the whole game."
It is helpful to step down a dimension, here from three to two. It <u>is</u> possible to dissect a square in the plane into a finite number $n(\geq 2)$ of unequal squares and it can be shown that at least 21 squares are needed to achieve this.
Now the bottom face A of the alleged dissected big cube is a square dissected into a finite number of squares, no two of the same size. (This number is finite (≥ 21) because the number of cubes is finite and they are of different sizes because the cubes are of different sizes.)
The smallest cube standing on A cannot be along the edge of the cube, that is, it is an internal cube. (This is because if the smallest square is along the edge, you cannot fit larger squares around it.) The smallest cube standing on A must thus be surrounded by larger cubes. Therefore, the top face of this smallest cube is surrounded by "walls" formed by the faces of the surrounding cubes. Therefore, the cubes standing on this top face B must give a square dissection of B and clearly B cannot be covered by a single cube.

But now we are back where we started, with a square dissection scenario. We take the smallest cube involved in the dissection of B and the process repeats. At every stage, we end up with a square dissection which requires smaller and smaller cubes to be added. The process never terminates, contradicting the fact that the dissection of the cube contains just a finite number of smaller cubes.

TOPICS FOR INVESTIGATION: Could you mimic the above proof to show that a rectangular box cannot be dissected into a finite number $n(\geq 2)$ of different cubes?

And how about higher dimensions? The problem is impossible in three dimensions, can you show it is impossible in $4, 5, 6, \ldots, n$ dimensions – that is not difficult. (The "face" of a four-dimensional hypercube would lead to a solution in 3D which we know does not exist!)

7.6

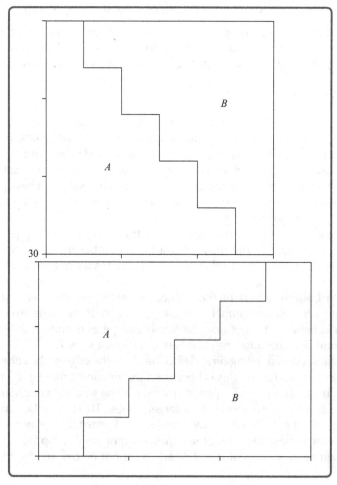

Rather amazingly, this problem can be solved in just one very lateral cut, which of course has to be the best possible result. It would work best for a plain carpet, but preserving a pattern would provide an interesting further constraint.

TOPICS FOR INVESTIGATION: What other integer dimensions could be used in this problem? This is one of the first step-dissections one encounters, but are other types of dissection possible?

Suppose you have a 12 × 12 carpet and you want to cut it into rectangular pieces to fit a 9 × 16 room. What is the best you can do now?

7.7

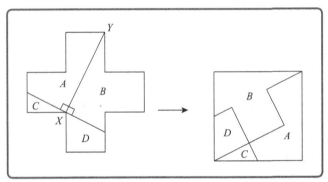

The lateral thought needed here is that the original diagram has area of five square units. If it is to be dissected and the pieces realigned to form a square, then that square must have side $\sqrt{5}$ units. We must provide right angles, so we can connect X and Y and draw a line perpendicular to it through X. The four pieces thus formed, A, B, C and D then magically fit together to form a square of area five square units.

TOPIC FOR INVESTIGATION: Can you see any connection between this dissection and any other dissections in this chapter?

7.8

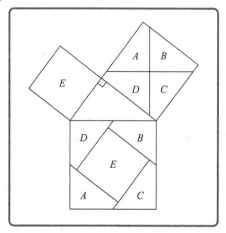

This beautiful dissection for the famous theorem of Pythagoras was discovered by the English mathematician Henry Perigal (1801–1898). The square on the hypotenuse is dissected into five pieces which then nicely cover the other two squares.

TOPICS FOR INVESTIGATION: Several other such dissections exist, some going back many hundreds of years. Can you experiment and find any of them?

Discuss the validity of dissection "proofs" in geometry.

What about the case where the right-angled triangle is isosceles? Show that you can find a nice dissection on a tiled floor.

7.9

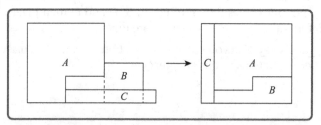

Again, we have a very elegant solution in three pieces, which surely must be the smallest possible.

TOPICS FOR INVESTIGATION: A good technique for generating mathematical problems is to add additional constraints. Here is one example:

Can you take a square and dissect it into <u>five</u> rectangular pieces which can then be reassembled to form <u>three</u> squares, no two of the same size?

[Hint: Look at the equation $1^2 + 4^2 + 8^2 = 9^2$.]

7.10

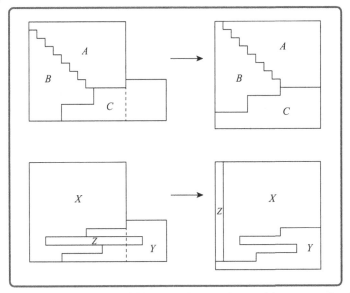

We present two very different solutions for problem 7.10, the first one is a step dissection. These can be found by experiment and a little lateral thinking, but they are by no means obvious and can be very time-consuming, and the amount of paper involved can put pressure on the rain forest!

TOPICS FOR INVESTIGATION: There is yet another solution, again a step dissection; can you find it?

And it looks very likely indeed that the desired result cannot be achieved with just two pieces, but how on earth could one prove this? (I can't!)

Matchsticks and Coins

8.1

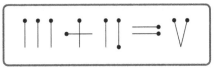

This is a fun chapter, but fun does not always mean easy!

The lateral trick is to take I from IV and use it to turn the – into a +.
This gives III + II = V

Topics for Investigation: Here is another nice solution:

$$VI - II = IV$$

Can you find a third solution?

8.2

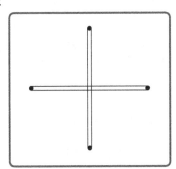

Move one match and observe the little square in the middle! The different drawing of the matchsticks should be a giveaway here.

Topics for Investigation: Is the vacant space given actually a square? You should have fun deciding what the answer is.
And while you are at it, what is a triangle? What is a circle? What is a disc?

DOI: 10.1201/9781003341468-20

8.3

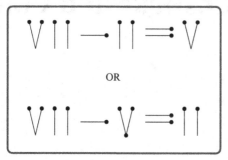

Here we cleverly transform one Roman numeral into another, equivalent to

$$7 - 2 = 5 \text{ or } 7 - 5 = 2$$

TOPICS FOR INVESTIGATION: What lateral transformations can you perform on L and M?

8.4

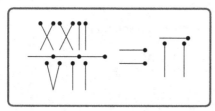

Quite lateral and easy if you spot it, but quite difficult if you don't!

$\frac{22}{7}$ is a well-known approximation for pi!

TOPICS FOR INVESTIGATION: The laterality of matchstick puzzles often consists of interpreting mathematical numbers, symbols or operations in a different way.

Can you think of any tricks that we have missed?

For example, in the incorrect equation

$$3.1 = 6,$$

can you move just one number to make the equation correct?

We cannot resist inserting this little beauty.

The equation $62 - 63 = 1$ is clearly not correct. Make it correct by

(i) Moving exactly one "object."

(ii) Moving exactly one digit.

These two solutions are pure laterality!

8.5

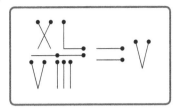

$$\frac{XL}{VIII} = V$$

can be represented in ordinary numerals as $\frac{40}{8} = 5$.

TOPICS FOR INVESTIGATION: What Roman numerals can be represented by matchsticks?

Well, we have I, V, X, L and M.

Can you make a matchstick puzzle involving all of these?

8.6

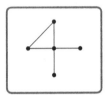

Surely you must agree that 4 is a square!

TOPICS FOR INVESTIGATION: Some people would regard this solution as "sneaky."

Discuss "sneaky" in mathematics.

Is "sneaky" equivalent to "lateral"?

Or is sneaky lateral's unacceptable cousin?

Would the following formulation of this problem have been better?

Make a square with just three matchsticks without bending or breaking any of them.

8.7

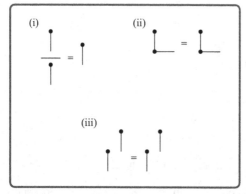

We give three solutions based on the following mathematical equations:

8.1 $1/1 = 1$ or $\frac{1}{1} = 1$

8.2 $L = L$ (Roman 50 = Roman 50)

8.3 $1^1 = 1^1$

TOPICS FOR INVESTIGATION: Surely you can find further solutions!

Would you accept $1 \neq 11$ as a valid solution?

And here is one of the very best and most lateral of all matchstick puzzles:

Make this equation correct by moving just one matchstick:

$$II = VI$$

It would be a pity to deprive you of the supreme pleasure of solving this very lateral problem by giving you the solution. Keep trying, keep trying, and that glorious moment will come!

[Would you accept the alternative solution II = XI?]

8.8

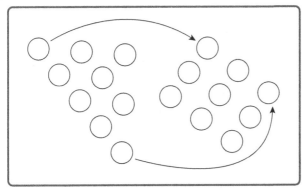

Tricky one but very satisfying!

A lateral thought – there are nine coins so the required square must be 3×3.

Topics for Investigation: Suppose the baseline has 7 coins. How many coins do you have to move to form a square?

Can you generalize the problem to where the baseline has $2n - 1$ coins, remembering the very beautiful fact that $1 + 3 + 5 + \ldots + (2n - 1) = n^2$?

8.9

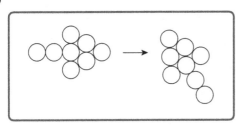

This one is almost too easy, but it does fool some people!

Topics for Investigation: Think about the meaning of "easy" in mathematics. Is one person's "easy" another person's "difficult"? Should these words be used at all? Should we not use just "true" and "false" instead? And how about "trivial"? And if a computer can be programmed to prove a mathematical result (and in some cases it can) can it decide if the problem is easy or difficult? Does it depend on the length of the proof?

But as one of my students once said, "A problem is easy if you can do it; difficult if you can't."

8.10

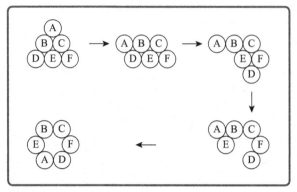

Step 1: Move *A* to touch *B* and *D*.

Step 2: Move *D* to touch *E* and *F*.

Step 3: Move *E* to touch *A* and *B*.

Step 4: Move *A* to touch *E* and *D*.

TOPICS FOR INVESTIGATION: This is more fun if you use real coins and experiment. Can you find a solution in less than four steps or show that four is the minimum number of steps required?

9 Logic

9.1

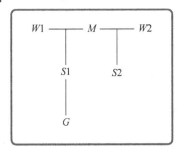

Two men are brothers if and only if they have the same father and the same mother. If two men have the same father and different mothers, they are half-brothers – but not brothers.

For simplicity, assume that man M marries woman $W1$ and they have a son $S1$ and no other children. Later on, $S1$ has a son G. Sadly, $W1$ dies and M marries $W2$ and they have a son $S2$ and no other children.

Now $S2$ can say of G, "Brothers and sisters I have none, but this man's father is my father's son."

TOPICS FOR INVESTIGATION: Do you regard this alternative solution as valid? What does your dictionary say? Are half-brothers brothers?

Do you have any difficulty with "story problems" in mathematics (as your author does)? Do they not almost always contain hidden and unstated assumptions, dependent on language and meaning?

9.2 Is "no" the answer to this question?

If "no" is the answer to this question, then the answer is "yes."

If "yes" is the answer to this question, then the answer is "no."

So we are in an impossible and contradictory situation. This is the epitome of Russell's paradox, invented by the philosopher-mathematician Bertrand Russell, and your author is very proud to have invented this version of the problem.

The only "solution" poor Russell could come up with was, essentially, don't ask such questions!

DOI: 10.1201/9781003341468-21

TOPICS FOR INVESTIGATION: In a little village with just one barber, the barber shaves every man who doesn't shave himself, but of course the barber does not shave anyone who shaves himself. Who shaves the barber? [There is a beautifully lateral solution to this version.]

Mathematically, we are in the realm of the set of all sets which do not contain themselves as members – does this set contain itself as a member or not?
[Recall that "the set of all phrases" is a phrase, but that the set of all carrots is not a carrot.]

9.3 If statement 1 is true, then the list contains exactly one false statement, so all of 2, 3, 4 and 5 are true except one. But any of 2, 3, 4 or 5 being true contradicts statement 1 being true.

If statement 2 is true, the list contains exactly two false statements, so three of the statements must be true. So 1 or 3 or 4 or 5 must be true, contradicting the fact that statement 2 is true.

If statement 3 is true, the list contains exactly 3 false statements, so two of the statements must be true, including 3.

So 1 or 2 or 4 or 5 must be true, contradicting the fact that statement 3 is true. So 3 is false. If 5 is true, all of the statements are false, contradicting the fact that 5 is true.

Finally, we see that statement 4 is true – the list contains exactly four false statements.

TOPICS FOR INVESTIGATION: Does this problem make your head swim a little? It should! What if we have 10 such statements? 1000? n, for any positive integer n? What is the solution then?

What if the word "exactly" is replaced by the words "at least" and "at most?"

9.4 The lateral step in this solution is that the shopkeeper's loss is the lady's gain. The shopkeeper next door comes out even in the transaction.

The lady gains $20 plus a pair of boots worth $30 making a total of $50. So the shopkeeper loses $50 in total.

TOPICS FOR INVESTIGATION: Try this puzzle on your friends and count the surprising number of different answers you are given. I have had four different answers!

The prices mentioned would indicate that this is a very old puzzle!

9.5 One of the most lateral activities one can indulge in is the finding of valid alternative solutions to mathematical problems, especially classical problems that everyone seems to know. Rather astonishingly, the solution here is due to my brilliant daughter Catherine, when she was only 5!
When I showed her the usual solution, she said, "Why not tie the duck to the back of the boat by a firm lead, and have it swim over and back?" First

take the fox over (one crossing), then return for the corn (two crossings) and finally take the corn over (three crossings). A bonus is the duck gets the much-needed exercise, swimming!

TOPICS FOR INVESTIGATION: Are there any other classical problems to which you can find a valid alternative solution? Recall our problem 9.2 for example.

9.6 It would be very difficult to beat this lateral solution.

You light both ends of the first rope and one end of the second. Then, when the first rope is completely burnt, you light the other end of the second rope. Exactly 45 minutes has elapsed when the second rope is totally consumed.

TOPICS FOR INVESTIGATION: Can you think of any other problems where a lateral solution involves turning something back to front or upside down?

9.7 If we divide 8 bottles of wine equally among 3 people, each person gets $2\frac{2}{3}$ bottles.

A has 3 bottles, so contributes $\frac{1}{3}$ bottles to C. B has 5 bottles, so contributes $2\frac{1}{3} = \frac{7}{3}$ bottles to C.

Now C pays $8 and B contributes 7 times as much wine to C as A does.

So B gets $7, while A gets $1.

Counterintuitive, eh?

TOPICS FOR INVESTIGATION: Look for other mathematical problems where the solution is surprising. How about this one:

A rope is tied tightly around the equator of the Earth. One meter is then added to the length of the rope, and it is then placed around the Earth at an equal distance from the equator everywhere! What is this distance?

Everyone finds the answer surprising!

9.8 How many letters are there in the correct answer to this problem?

Several lateral solutions are possible.

1. THE CORRECT ANSWER TO THIS PROBLEM has 29 letters.
2. 0 – no letters.
3. Zero – four letters

TOPICS FOR INVESTIGATION: Can you think of any other solutions?

Think about the uniqueness of a solution to a mathematical problem. Is a problem "better" if it has a unique solution? Are real life problems more likely or less likely to have a unique solution?

Do you know of anyone with precisely the same name as yourself? (First name and second name, and spelling.) Medical workers now use the patient's date of birth as an identification. What are the chances of two people having the same name and the same date of birth?

9.9 You say to him, "If I were to ask a member of the other tribe, which was the correct way to the city, which way would he say?"

Whatever road he points to, you take the other one!

If he is a truth-teller, he will point to the wrong way, knowing that a member of the other tribe would lie.

If he is a liar, he would also point to the wrong way, knowing that a member of the other tribe would tell the truth.
So take the other road in each case!

TOPICS FOR INVESTIGATION: The question asked is a complex one – "if I were to ask, ..., what would he say?" Is it possible to solve the problem by asking a simple question?

A very lateral solution is to ask, "Did you know they are giving out free beer in the city today?" and then follow him both along the correct road!

And we must bear in mind the definition of a "liar." If it is a person who is always determined to deceive and mislead you, he may sometimes tell the truth to accomplish this! And if the man is a politician, the problem is hopeless.

9.10 The statement you should make is "I will be hanged."

If this statement is true, then you will be shot, so the statement is false.

If this statement is false, then you will be hanged, so the statement is true.

We are now back in the impossible situation of problem 9.2, so the chief with his doctorate in logic (supervisor Bertrand Russell) allows me to go free.
Who said logic was not useful?

TOPICS FOR INVESTIGATION: Your author is very interested in the connections between logic and, mathematics and humor. So as a relaxing break, here are two joke riddles for you:

1. What does a person with a degree in Philosophy say to you?

2. What does a person with a Ph.D. in Philosophy say to you?

ANSWERS

1. Do you want fries with that?

2. Why do you want fries with that?

Maxima and Minima

10.1 Good lateral thinking does not preclude experimentation! In fact many people believe that mathematics is an experimental science. Never forget this!

How about $50 + 50 = 100$ giving a product of 2500? Clearly, this is not the best we can do, because for example $20 + 30 + 50 = 100$ gives a bigger product of 3000.

In fact, if in any sum say 10 occurs, we can achieve the same sum by writing $10 = 3 + 7$ and $3 \times 7 = 21 > 10$. In general, if $r > 3$, $k = k + (r - k)$ and $k(r - k) > k$, whenever $r > k$.

So we can decrease each component of the sum in turn and increase, or certainly not decrease, the product. For example,

$$7 = 4 + 3 \text{ and } 4 \times 3 = 12 > 7$$
$$6 = 3 + 3 \text{ and } 3 \times 3 = 9 > 6$$
$$5 = 3 + 2 \text{ and } 3 \times 2 = 6 > 5$$
$$4 = 2 + 2 \text{ and } 2 \times 2 = 4 = 4.$$

Thus, we can reduce the problem to a combination of 2s and 3s.

Incidentally, involving a 1 is a waste of time, because it increases the sum but keeps the product the same.

So how many 2s and how many 3s should we use? Notice that $2 \times 2 \times 2 = 8$ while $3 \times 3 = 9$, while $2 + 2 + 2 = 6 = 3 + 3$. This tells us that we should maximize the number of 3s and make the rest equal to 2.

For a sum of 100, the maximum product is $3^{33} \cdot 2$, which you need not work out!

TOPICS FOR INVESTIGATION: What if the sum is 1000? 1,000,000? n?

Do you know of any number between 2 and 3 (and closer to 3) which might have a bearing on the given solution?

For readers who have studied the sublime topic of group theory, the answer may be interpreted as the order of the abelian subgroup of maximum size in the symmetric group S_{100}.

DOI: 10.1201/9781003341468-22

10.2

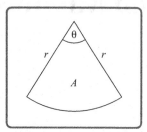

We have $2r + r\theta = l$, so $\theta = (1/r)(l - 2r)$. Now area of the sector $A = (1/2)r^2\theta = (1/2)r^2(1/r)(l - 2r) = (1/2)rl - r^2$. Now this can be written as $(1/16)l^2 - (r - (1/4)l)^2$. Since $(r - (1/4)l)^2 \geq 0$ always, so the maximum value of A is $(1/16)l^2$, achieved when $r = (1/4)l$. Then $\theta = (l - 2r)r = 2$ (radians) which is about $114°35'$.

Topics for Investigation: Notice that in the case of maximum area, the straight part of the wire has the same length as the curved part. Why do you think that is?

Could you formulate and solve a three-dimensional version of this problem?

Again, compare the calculus solution with the given solution. Which do you prefer?

10.3 This is a geometric problem which looks non-trivial – but think laterally in another box. How about a trigonometric approach?
Let A be the angle opposite the third and variable side of the triangle.

Then the area of the triangle is $(1/2)(10)(10) \sin A = 50 \sin A$.
Now A is the only variable we have control over and $\sin A \leq 1$, attaining its maximum of 1 here when $A = 90°$.
So for maximum area, the triangle is right-angled of area 50 square units and the length of the third side is $10\sqrt{2} = 14.14\ldots$

Topics for Investigation: Try this problem using Heron's famous formula for the area of a triangle $\Delta = \sqrt{s(s - a)(s - b)(s - c)}$.
If the unknown side is $2x$, $s = (1/2)(10 + 10 + 2x) = 10 + x$; $(s - a) = x$, $(s - b) = x$ and $(s - c) = 10 - x$. Then the area is $x\sqrt{100 - x^2}$.
How does one maximize that? Well, you could use calculus or use the observation that Δ is a maximum whenever Δ^2 is a maximum and continue to get the above solution. But is not the lateral solution much more satisfying?

Actually, there is something very subtle going on here that is of interest. Usually maximization and minimization problems for a triangle finish with the phrase, "which happens if and only if the triangle is equilateral" but here that is not the case; the maximum occurs if and only if the triangle is right-angled isosceles. Can you think of any other maximization or minimization problems for a triangle where this is the case?

10.4 The biggest integer one can make with just three digits appears to be $9^{(9^9)}$. Can you do any better?

Topics for Investigation: Note that $9^{(9^9)}$ is not the same as $(9^9)^9$. What is the value of each of these numbers?

What is the biggest number you can make with four digits? Five digits? n digits?

10.5 If x is a positive real number, then $(\sqrt{x} - \frac{1}{\sqrt{x}})^2 \geq 0$ or $(\sqrt{x})^2 + \frac{1}{(\sqrt{x})^2} - 2\sqrt{x} \cdot \frac{1}{\sqrt{x}} \geq 0$. So $x + \frac{1}{x} - 2 \geq 0$ or $x + \frac{1}{x} \geq 2$. If $\sqrt{x} - \frac{1}{\sqrt{x}} = 0$, then $\sqrt{x} = \frac{1}{\sqrt{x}}$ or $x = 1$ and $x + \frac{1}{x} = 2$ as required.

Topics for Investigation: How does this solution compare with a solution using calculus?

Can you find the minimum value of $ax + b/x$ for positive x? Of $ax^2 + b/x^2$, for positive a and b?

How about using the AM-GM inequality?

10.6

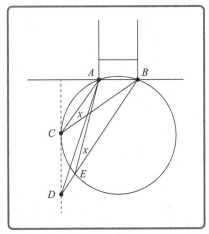

Draw the (unique) circle which goes through A and B, the bases of the goal posts, and touches the dotted line at some point C. C is then the optimal point at which to place the ball to attempt a kick at goal. This is because the $\angle ACB$ is maximal for any point C on the dotted line.

We know this because let D be any other point on the dotted line and let BD meet the circle at E. Join AE. Then, because of the well-known theorem that angles on the same chord AB are equal, we have $\angle AEB = \angle ACB$. But by another theorem, $\angle AEB$ is greater than $\angle ADB$, since the exterior angle of a triangle is greater than the interior remote angle. Thus $\angle ACB$ is greater than $\angle ADB$ for any point D which is not C.

This method is used to pick a most favorable conversion point after a try in Rugby Union.

TOPICS FOR INVESTIGATION: The original problem is three-dimensional. How valid is the two-dimensional solution we have given?

10.7 We seek the minimum value of $f(x) = ax^2 + bx + c$, for $a > 0$, without the use of calculus.

Since $a > 0$, $f(x)$ is a minimum if and only if $(4a)f(x) = 4a^2x^2 + 4abx + 4ac$ is a minimum.

Now

$$4a^2x^2 + 4abx + 4ac = 4a^2x^2 + 4abx + b^2 + (4ac - b^2)$$
$$= (2ax + b)^2 + (4ac - b^2).$$

Now $(2ax + b)^2 \geq 0$ always and $(2ax + b)^2 = 0$ if and only if $2ax + b = 0$, that is, $x = (-b)/(2a)$. So the minimum value is $\frac{4ac-b^2}{4a}$.

TOPICS FOR INVESTIGATION: Where have we used the fact that $a > 0$? What happens if $a < 0$? $a = 0$? Illustrate graphically.

If you have studied calculus, solve the problem using calculus. Which do you think is the better method, and why?

10.8 The first and sensible thing to do is to consult calendars and see what the answer is likely to be (MIAES!).

It quickly becomes obvious that there is a difference between regular years and leap years, so these must be treated separately. It is also true that the 13th falls on a Friday if and only if the first day of the month is a Sunday. For a regular (non-leap) year the months which always begin on the same day of the week fall into groups. These are

JANUARY OCTOBER	MAY	AUGUST	FEBRUARY MARCH NOVEMBER	JUNE	SEPTEMBER DECEMBER	APRIL JULY

Thus there are a maximum of three Friday the 13th and a minimum of one in any regular year.

In the case of a leap year, the pattern is

JANUARY APRIL JULY	OCTOBER	MAY	FEBRUARY AUGUST	MARCH NOVEMBER	JUNE	SEPTEMBER DECEMBER

Again, we have that there are a maximum of three and a minimum of one Friday the 13ths in any leap year.

TopIC FOR INVESTIGATION: Produce several actual regular years and leap years where the maximum of three and the minimum of one occur. What if we do not confine ourselves to a calendar year but consider just twelve consecutive months instead? What are the maximum and minimum number of Friday the thirteenths now?

And here is an extraordinary fact that you might like to prove:

Overall, the 13th of the month is more likely to fall on a Friday than any other day of the week!

And just for completeness, here are two other calendar problems for your entertainment.

1. A month is called "bad" if it has five Mondays. What is the maximum and minimum number of "bad" months in a calendar year? In a 12 month period?

2. How many different calendars does one need to cover all years?

NOTE: Some readers who have studied abstract algebra may recognize equivalence classes in our solutions – two months are equivalent if they begin on the same day of the week.

10.9 Look, the differential calculus is probably the most useful and powerful mathematical tool ever invented, so let's use it here, with some nice lateral touches.

If A is the total surface area of a closed right circular cylinder of radius r and height h, we have $A = 2\pi r^2 + 2\pi r h$.

Now let $B = A/2\pi$, so since A is constant, B is constant too, so let's work with B.

Thus $h = B/r - r$.

Then $V = \pi r^2 h = \pi r^2 \left[\frac{B}{r} - r\right] = \pi \left[Br - r^3\right]$.

Again, we can drop the constant π.

$\frac{dV}{dr} = B - 3r^2 = 0$ for a maximum.

[Note $\frac{d^2V}{dr^2} = -6r < 0$.]

So $B = 3r^2$ and $h = B/r - r = 3r^2/r - r = 3r - r = 2r$.

So for maximum volume for a given surface area, we must have $h = 2r$, that is, total height equals total diameter.

TOPICS FOR INVESTIGATION: How about a cylinder open at one end?

These problems clearly have big implications in the canning industry.

How about the reverse situation where we have a fixed volume for a right circular cylinder (open or closed) and seek the minimum surface area?

10.10 Think laterally! All the really big integers involve millions, billions, trillions, zillions and the like, and these end in ION with the ON out of alphabetical order.

Even a thousand or a hundred present will clearly not work. Ninety has NI, eighty has IG, seventy has VE, sixty has XT and fifty has IF. But FORTY fits the bill and turns out to be the correct (and only) answer!

TOPICS FOR INVESTIGATION: What is the largest integer written in the English language, where the letters are in reverse alphabetical order?

How about other languages such as French, German, Spanish and Italian?

Calculus and Analysis

11.1 We seek $\underset{n\to\infty}{Lt}\ \frac{2^n}{n!}$.

Again, we jump laterally, not just outside the box, but into another box – the theory of infinite series.

Consider $\sum_{n=1}^{\infty}\frac{2^n}{n!}$. Using the well-known ratio test for convergence, we examine

$$\frac{a_{n+1}}{a_n} = \frac{2^{n+1}}{(n+1)!} \cdot \frac{n!}{2^n} = \frac{2}{n+1}.$$

Now $\underset{n\to\infty}{Lt}\left|\frac{a_{n+1}}{a_n}\right| = \underset{n\to\infty}{Lt}\ \frac{2}{n+1} = 0$, so this series converges. But another well-known result tells us that for a convergent series, the nth term must tend to zero as n tends to infinity, so $\underset{n\to\infty}{Lt}\ \frac{2^n}{n!} = 0$. Lateral magic!

Topics for Investigation: Use this technique to show that $\underset{n\to\infty}{Lt}\ \frac{n!}{n^n} = 0$. It works beautifully if you know a nice limit involving e.

11.2

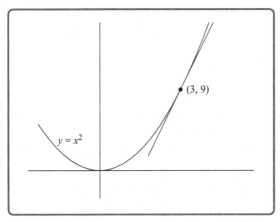

A line is a tangent to a curve if it has "double contact" with the curve at a point $x = a$. Another way of putting this, at least for polynomial functions, is that there is a repeated factor $(x - a)^2$.

DOI: 10.1201/9781003341468-23

Let the tangent line to the curve $y = x^2$ at the point $(3, 9)$ be $y = mx + c$, and since this line passes through $(3, 9)$, we have $9 = 3m + c$ so that $c = 9 - 3m$ and the tangent has equation $\underline{y = mx + (9 - 3m)}$.

At the point of intersection of the curve and the tangent line we have $y = x^2$, so $x^2 = mx + (9 - 3m)$ or $x^2 - mx + (3m - 9) = 0$.
For this quadratic to have a repeated root, it must be a perfect square, so we get $m^2 = 4 \cdot 1 \cdot (3m - 9)$ or $m^2 - 12m + 36 = 0$ which is $(m - 6)^2 = 0$, giving $m = 6$.

Thus the equation of the tangent is $y = 6x - 9$, exactly the same result as we get from calculus!

TOPIC FOR INVESTIGATION: Can you find the equation of the tangent to the curve $y = x^3$ at the point $(2, 8)$ by the above method?
How about $y = \sin x$ at $(\pi/4, 1/\sqrt{2})$?

Discuss whether we really need calculus for finding the equations of tangents.

11.3 $I = \int \sec x \, dx$.

Actually, there are many different ways of evaluating I, but this surely is the most lateral way.

$$I = \int \sec x \, dx = \int \frac{\sec x (\sec x + \tan x)}{\sec x + \tan x} \, dx$$

[Why are we multiplying above and below by $(\sec x + \tan x)$? Be patient and you will see that it works!]

$$= \int \frac{\sec^2 x + \sec x \tan x}{\sec x + \tan x} \, dx = \int \frac{\sec x \tan x + \sec^2 x}{\sec x + \tan x} \, dx.$$

Now recall that $\frac{d(\sec x)}{dx} = \sec x + \tan x$ and $\frac{d(\tan x)}{dx} = \sec^2 x$.

So $I = \int \frac{d(\sec x + \tan x)}{\sec x + \tan x} = \ln |\sec x + \tan x| + c$.

TOPIC FOR INVESTIGATION: In what other ways can you evaluate $\int \sec x \, dx$? How do they compare with the above?

Can you think of any other lateral methods of evaluating antiderivatives or integrals?

How about $I_1 = \int \sin^2 x \, dx$ and $I_2 = \int \cos^2 x \, dx$? Look at $I_1 + I_2$ and $I_2 - I_1$.
Try also $I_3 = \int \tan^2 x \, dx$.

11.4 This is one of the most famous and most useful results in mathematics and there are literally dozens of different proofs. This one in our opinion is the most lateral.

In $(1/n)(x_1 + x_2 + \ldots + x_n) \geq \sqrt[n]{x_1 x_2 \ldots x_n}$, the left hand side involves a sum and the right hand side involves a product. Where else do we find interaction between sums and products? Well, powers and logs give us an example.

It is not difficult to prove that $x \leq e^{x-1}$ (∗) for all real numbers x, with equality if and only if $x = 1$. One can use calculus or the exponential series – we leave this as an exercise to the reader so as not to distract from the beautiful laterality of our solution.

Let our set of non-negative real numbers be $\{x_1, x_2, \ldots, x_n\}$. If all of these numbers are zero then the AM-GM inequality clearly holds, actually with equality. Thus we may assume in what follows that their arithmetic mean $(1/n)(x_1 + x_2 + \ldots + x_n) = a > 0$.

Applying (∗) above n times, we obtain

$$\frac{x_1}{a} \cdot \frac{x_2}{a} \cdot \ldots \cdot \frac{x_n}{a} \leq e^{\frac{x_1}{a}-1} \cdot e^{\frac{x_2}{a}-1} \cdot \ldots \cdot e^{\frac{x_n}{a}-1}$$

$$= \exp\left(\frac{x_1}{a} - 1 + \frac{x_2}{a} - 1 + \ldots + \frac{x_n}{a} - 1\right)$$

with equality if and only if $x_i = a$, $1 \leq i \leq n$. Now,

$$\frac{x_1}{a} - 1 + \frac{x_2}{a} - 1 + \ldots + \frac{x_n}{a} - 1 = \frac{x_1 + x_2 + \ldots + x_n}{a} - n$$

$$= \frac{na}{a} - n$$

$$= n - n$$

$$= 0.$$

So

$$\frac{x_1 x_2 \ldots x_n}{a^n} \leq e^0 = 1.$$

Finally, we get $x_1 x_2 \ldots x_n \leq a^n$ or $\sqrt[n]{x_1 x_2 \ldots x_n} \leq (1/n)(x_1 + x_2 + \ldots + x_n)$, which is the famous AM-GM inequality. With a bit more work, we can show that equality holds if and only if $x_1 = x_2 = \ldots = x_n$.

It is believed that this beautiful proof was discovered by the great mathematician George Pólya.

TOPICS FOR INVESTIGATION: Seek out some of the many other proofs of the AM-GM inequality. Do you agree that the one we have first given is the most lateral one or can you make a good case for another proof?

In addition, you now have a golden opportunity to look at the many wonderful applications of the AM-GM inequality in mathematics. For example, show that $[(n + 1)/2]^n \geq n!$ for each natural number n, with equality if and only if $n = 1$.

And a geometric exercise: Show that $s^2 \geq 4\Delta$, where $s = (1/2)(a + b + c)$ is the semiperimeter and Δ its area, but that equality <u>never</u> holds. Can you improve this result by considering $n = 3$ rather than $n = 4$?

11.5 One of the most significant descriptions of lateral thinking is of looking at a problem from a different point of view and in this example that is literally true.

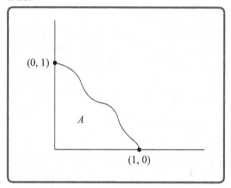

Consider the plane curve given by the equation

$$x^7 + y^4 = 1.$$

We really don't need to know what it looks like exactly, but it will be something like in the diagram, passing through $(0, 1)$ and $(1, 0)$.

We have $y = \sqrt[4]{1 - x^7}$, so by integral calculus, the area A is given by

$$A = \int_0^1 \sqrt[4]{1 - x^7}\, dx$$

However, rotating the page through $90°$, we have $x = \sqrt[7]{1 - y^4}$, and so $A = \int_0^1 \sqrt[7]{1 - y^4}\, dy$ also. So the two numbers are equal!

<u>TOPIC FOR INVESTIGATION</u>: Can you generalize this question by considering $x^n + y^m = 1$?

11.6 It turns out that neither of the given integrals is *easy* (or even possible) to evaluate individually, but that their *difference* is easy to evaluate.

The lateral trick is that x and y are "dummy" variables and each may be substituted for the other – in a *definite* integral, but not in an *indefinite* integral!

Let $I_1 = \int_{\pi/4}^{\pi/2} \left(\frac{\sin x}{x}\right)^2 dx; I_2 = \int_{\pi/4}^{\pi/2} \frac{\sin 2x}{x} dx.$

Then $I_1 - I_2 = \int_{\pi/4}^{\pi/2} \frac{\sin^2 x - x \sin 2x}{x^2} dx = \int_{\pi/4}^{\pi/2} \frac{d}{dx}\left(\frac{-\sin^2 x}{x}\right) dx.$

[Note $\frac{d}{dx}\left(\frac{-\sin^2 x}{x}\right) = \frac{\sin^2 x - x \sin 2x}{x^2}$, using $d\left(\frac{u(x)}{v(x)}\right)$.]

$= \left[\frac{-\sin^2 x}{x}\right]_{\pi/4}^{\pi/2} = \ldots$

We leave the messy but easy evaluations of the numbers as an exercise.

TOPIC FOR INVESTIGATION: Can you manufacture an example where $\sum_{r=1}^{\infty} f(r)$ and $\sum_{r=1}^{\infty} g(r)$ are either difficult or impossible to evaluate (or may not even exist!) but their difference can be evaluated?

Can you find $\lim_{x \to \pi/2}[\sec x - \tan x]$?

11.7 One can of course solve this problem directly, but the working becomes a little messy and it is easy to make a mistake.

And what if you were asked for the 100th derivative or the 1000th derivative?

There is an old lateral trick in mathematics which can be used in many contexts. It is called "partial fractions" and we can use it here. It is essentially a divide and conquer technique.

We write $\frac{1}{1-x^2} = \frac{A}{1+x} + \frac{B}{1-x}$, for some constants A and B.

Then $\frac{1}{1-x^2} = \frac{A(1-x)+B(1+x)}{(1+x)(1-x)} = \frac{x(B-A)+(A+B)}{1-x^2}$. So $1 = x(B - A) + (A + B)$ or $B - A = D$ and $A + B = 1$, giving $A = B = (1/2)$ and $\frac{1}{(1-x^2)} = \frac{1}{2(1+x)} + \frac{1}{2(1-x)}$.

[This should be checked carefully because we have assumed that A and B exist – in general, they may not.]

So

$$\frac{d^{20}}{dx^{20}}\left(\frac{1}{1-x^2}\right) = \frac{20!}{2}\left[\frac{1}{(1+x)^{21}} - \frac{1}{(1-x)^{21}}\right].$$

TOPIC FOR INVESTIGATION: Many books on algebra have a whole chapter on partial fractions. Find one and read it.

Do you think you can find the nth derivative of $1/(1 - x^2)$?

Can you find an example where you assume that a partial fraction decomposition of a certain kind exists but in fact it does not?

11.8 Let $I = \int \frac{dx}{\sin x + \cos x}$.

There are several ways of finding this integral or antiderivative. One is by making the substitution $t = \tan(x/2)$. Then $\sin x = (1 - t^2)(1 + t^2)$; $\cos x = 2t/(1 + t^2)$ and $dx = 2dt/(1 + t^2)$ giving $\int dt/(t^2 + 2t - 1)$ which is easily evaluated.

A more lateral way is to multiply above and below by $1/\sqrt{2} = \cos(\pi/4) = \sin(\pi/4)$. Then

$$I = \int \frac{\frac{1}{\sqrt{2}}dx}{\frac{1}{\sqrt{2}}\sin x + \frac{1}{\sqrt{2}}\cos x}$$

$$= \frac{1}{\sqrt{2}} \int \frac{dx}{\sin x \sin(\pi/4) + \cos x \sin(\pi/4)}$$

$$= \frac{1}{\sqrt{2}} \int \frac{dx}{\cos(x - (\pi/4))}$$

$$= \frac{1}{\sqrt{2}} \int \sec\left(x - \frac{\pi}{4}\right)dx$$

$$= \frac{1}{\sqrt{2}} \ln\left|\sec\left(x - \frac{\pi}{4}\right) + \tan\left(x - \frac{\pi}{4}\right)\right| + c.$$

The various answers you get may look different, but if you have worked correctly, any two will differ by an arbitrary constant.

But there is a truly lateral (= sneaky?) way of proceeding. Look up the answer for $\int \frac{dx}{\sin x + \cos x}$ in a table of integrals or on one of the many websites giving this information. Suppose the answer given is $f(x)$. Now differentiate the answer to get $1/(\sin x + \cos x)$.

This is perfectly legitimate because the definition of an antiderivative is a function which when differentiated leads to the given function. And if you have qualms of conscience about this procedure, how do you know, for example, that $\int \cos x\,dx = \sin x + c$? Only because you know that $\frac{d}{dx}(\sin x) = \cos x$!

TOPIC FOR DISCUSSION: Would the above justification convince a teacher or perhaps, more importantly, a mathematical examiner?

Anyway, what are we being asked to do in mathematics? To solve problems? To give answers? To show how we solved a problem? Or what?

11.9 The exponential function e^x, or exp x, is one of the most important functions in mathematics. It can be approached from many different angles, all of which are significant and useful.

(i) e^x is the unique non-trivial function whose derivative is equal to itself, that is,

$$\frac{d(e^x)}{dx} = e^x.$$

(ii) Power series definition: $e^x = 1 + x + \frac{x^2}{2!} + \frac{x^3}{3!} + \ldots = \sum_{n=0}^{\infty} \frac{x^n}{n!}$.

(iii) Limit definition: $e^x = \underset{n \to \infty}{Lt} \left(1 + \frac{x}{n}\right)^n$.

Graphically e^x looks like this, familiar to many people as exponential growth.

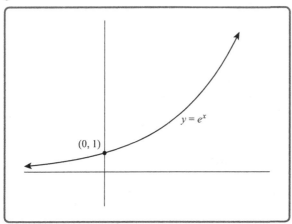

But however you define e^x, it is clear that $e^x > 0$, for all real numbers x. This gives rise to one of the sweetest lateral tricks in mathematics.

$\frac{d}{dx}(e^x) = e^x > 0$ for all x.

So e^x is always increasing for all x. Thus if $e^a = e^b$, we cannot have $a < b$ or $b < a$, so $a = b$.

TOPIC FOR INVESTIGATION: What can you say if $f'(x) > 0$ for all x, with regard to $f(a)$ and $f(x)$?

How about $\log_e a = \log_e b$?

$a^2 = b^2$, if a and b are both positive?

Finally, if $x^3 = y^3$ is $x = y$ always?

11.10 A vessel with infinite surface area but with a finite volume? It seems as if such an object could not exist, but there is an example called Gabriel's Horn or Torricelli's Trumpet, invented by the great Evangelista Torricelli (1608–1674). He was a student of Galileo and is believed to have invented the barometer.

Consider the curve $y = 1/x$, for $x \geq 1$.

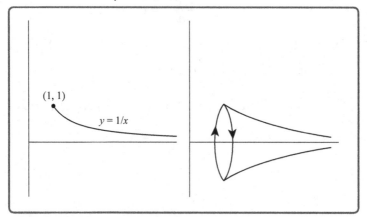

Now rotate this curve about the x-axis to form a trumpet-like solid, continuing indefinitely.

The volume of revolution is given by

$$V = \int_1^\infty \pi y^2 dx = \pi \int_1^\infty \frac{dx}{x^2} = \pi \left[-\frac{1}{x} \right]_1^\infty = \pi,$$

which is a finite volume.

The surface area of the solid is given by $\int_1^\infty 2\pi y \sqrt{1 + \left(\frac{dy}{dx}\right)^2}\, dx$. This integral is difficult to evaluate, but it is greater than

$$2\pi \int_1^\infty y\, dx = 2\pi \int_1^\infty \frac{dx}{x}$$
$$= 2\pi[\log_e \infty - 0] = \infty,$$

since $\log_e x \to \infty$ as $x \to \infty$. So the surface area is infinite!

TOPIC FOR INVESTIGATION: What happens if you fill a solid such as this with an appropriate finite amount of paint? It fills the solid completely, but does it not also cover the infinite surface area? How do you explain this? Look up the Painter's Paradox on the web for an explanation.

And decide if the opposite situation can occur, that is, a solid with a finite surface area but with an infinite volume.

Finally, investigate if you can make a model from putty or Plasticine which might make some sense of Gabriel's Horn.

12 A Mixed Bag

12.1 Look at the number of letters in each word in the sentence

How	I	wish	I	could	remember	x	easily	today
3	1	4	1	5	9	?	6	5

This should look familiar – they are the digits of π. So $x = \pi =$ pi, which has two digits –

$$\pi = 3.14159265\ldots$$

TOPIC FOR INVESTIGATION:

Can you continue this mnemonic with words of suitable length?

What other useful mnemonics in mathematics do you know? STCA? SO-HCAHTOA?

12.2 If we consider only those lines which are neither parallel nor perpendicular to the axes, we have

$$y = m_1 x + c \, (l_1); y = m_2 x + c_2 \, (l_2); y = m_3 x + c_3 \, (l_3)$$

where the values of c_1, c_2 and c_3 are immaterial.

l_1 is perpendicular to l_3, so $m_1 m_3 = -1$
l_2 is perpendicular to l_3, so $m_2 m_3 = -1$

So $m_1 m_3 = m_2 m_3$ and since $m_3 \neq 0$, we have $m_1 = m_2$ and so l_1 is parallel to l_2.

TOPIC FOR INVESTIGATION: Can we cover the cases where the lines are parallel or perpendicular to the axes?
Do you believe that $0 \cdot \infty = -1$?

René Descartes revolutionized mathematics by effectively turning geometry into algebra. Can you find other examples like the above?

And can you reverse the process by looking at $(a + b)^2 = a^2 + 2ab + b^2$ geometrically?

DOI: 10.1201/9781003341468-24

12.3 Let X be an $n \times n$ matrix whose entries are real or complex numbers. If $X^2 = X$, we say that X is idempotent.

Now the only $n \times n$ matrix which is idempotent and invertible is the identity matrix I_n. This is because $X^2 = X$ implies $X^{-1}X^2 = X^{-1}X$, so $X = I_n$.

Now suppose that $AB = I_n$, so that $0 \neq 1 = \det I_n = \det(AB) = \det A \cdot \det B = \det B \cdot \det A = \det(BA)$, so $\det(BA) \neq 0$ and BA is invertible since $\det(BA) \neq 0$.

Also $(BA)(BA) = B(AB)A = BI_nA = BA$ so that BA is idempotent and invertible, so $BA = I_n$.

This is a very beautiful trick and quite lateral.

Topic for Investigation: Show that the above result is not true for infinite matrices, that is, find infinite matrices A and B with $AB = I_\infty$ (1 everywhere down the infinite diagonal and 0 elsewhere) but $BA \neq I_\infty$.

12.4 A little thought suggests that $(x^2 - y^2)^2$ and $(z^2 - w^2)^2$ should be involved in the answer, and to cancel the terms we do not want, $(xy - zw)^2$ should be involved also. The lateral trick is to involve $(xy - zw)^2$ twice to complete the job.

It is then easy to conclude that $x^4 + y^4 + z^4 + w^4 - 4xyzw = (x^2 - y^2)^2 + (z^2 - w^2)^2 + (xy - zw)^2 + (xy - zw)^2$, easily verified by multiplying out the expressions involved.

If x, y, z and w are all non-negative real numbers, since the square of a non-negative real number is always non-negative, we conclude that $x^4 + y^4 + z^4 + w^4 \geq 4xyzw$.

It is an easy exercise to show that equality occurs if and only if $x = y = z = w$.

Topics for Investigation: Can you show that $x^4 + y^4 + z^4 + w^4 - 4xyzw$ does not factor algebraically in any non-trivial way?

What can we say about $x^2 + y^2 - 2xy$? $x^3 + y^3 + z^3 - 3xyz$? And in general, $x_1^n + x_2^n + \ldots + x_r^n - nx_1x_2 \ldots x_n$?

12.5 "Completing the square" to solve a quadratic equation is the method normally used in textbooks. However, this is a little messy algebraically, so first here is a more elegant version, with less fractions involved.

Let $ax^2 + bx + c = 0$, where $a \neq 0$, b and c are real or complex numbers. Since $a \neq 0$, the equation is equivalent to the equation

$$4a^2x^2 + 4abx + 4ac = 0$$

or (and this is the lateral bit)

$$4a^2x^2 + 4abx + b^2 = b^2 - 4ac.$$

This is $(2ax + b)^2 = b^2 - 4ac$,
or $2ax + b = \pm \sqrt{b^2 - 4ac}$,
or $2ax = -b \pm \sqrt{b^2 - 4ac}$,
and $x = \frac{-b \pm \sqrt{b^2 - 4ac}}{2a}$.

For completeness, these two solutions should be verified.

However, one can approach the problem from a completely different point of view. A quadratic equation has at most two solutions and these of course may be equal. (It is a nice exercise to show that a quadratic equation cannot have more than two solutions – do try it.)

Let the solutions of $ax^2 + bx + c = 0$ be p and q.
Then $(x - p)(x - q) = 0 = x^2 + (b/a)x + (c/a)$ so that $p + q = -(b/a)$ and $pq = (c/a)$.
Then $(p + q)^2 = p^2 + 2pq + q^2 = b^2/a^2$ and $-4pq = -4(c/a) = -(4ac/a^2)$.
So

$$(p - q)^2 = (p + q)^2 - 4pq = (b^2/a^2) - (4ac)/a^2$$
$$= (b^2 - 4ac)/a^2,$$

and $p - q = \pm(\sqrt{b^2 - 4ac}/a)$, $p + q = -(b/a)$.
Solving these equations for p and q, we get

$$p = \frac{-b + \sqrt{b^2 - 4ac}}{2a}, \quad q = \frac{-b - \sqrt{b^2 - 4ac}}{2a}.$$

TOPICS FOR INVESTIGATION: It took mathematicians many hundreds of years to go one step further and solve a cubic equation $ax^3 + bx^2 + cx + d = 0$, where $a \neq 0$. Go online or read any advanced textbook on algebra for details, which are quite demanding. The problem can also be solved for $n = 4$, quartic equations, but there is a disappointing surprise for $n \geq 5$.

Finally, look up the esoteric topic of quaternions. Here the equation $x^2 + 1 = 0$ has three solutions!

12.6 Evaluation of the definite integral

$$I = \int_0^1 \frac{x^4(1-x)^4}{1+x^2}\,dx$$

is quite straightforward. It is the conclusion that is startling! By long division, we get

$$\frac{x^4(1-x)^4}{1+x^2} = x^6 - 4x^5 + 5x^4 - 4x^2 + 4 - \frac{4}{1+x^2}.$$

Integrating term by term and evaluating between the limits 0 and 1 and remembering that $\int \frac{dx}{1+x^2} = \tan^{-1} x$, we get

$$I = \frac{1}{7} - \frac{4}{6} + \frac{5}{5} - \frac{4}{3} + 4 - 4\left(\frac{\pi}{4}\right) = 22/7 - \pi.$$

But the integrand is positive for all x in $(0, 1)$ so $22/7 > \pi$.
Lateral magic!

TOPIC FOR INVESTIGATION: Find a lower bound for π by observing that $x(1-x) \le (1/4)$ for all $0 \le x \le 1$. You should get $0.999 < 7\pi/22 < 1$.

Can you obtain closer and closer estimates for π by looking at different integrals of this kind?

For example, $\int_0^1 \frac{x^{2n}(1-x)^{2n}}{1+x^2}\,dx$, where n is a natural number.

12.7 $\sqrt{5} - 2 = \sqrt{5} - 2$. Now,

$$\frac{1}{\sqrt{5} + 2} = \frac{(\sqrt{5} - 2)}{(\sqrt{5} + 2)(\sqrt{5} - 2)} = \frac{\sqrt{5} - 2}{(\sqrt{5})^2 - 2^2} = \sqrt{5} - 2.$$

Now, $\frac{1}{\sqrt{5}+2}$ is $\sqrt{5} - 2$ with just two strokes added, so we are done!

TOPIC FOR INVESTIGATION: What part do tricks and puzzles like this play in the understanding and development of mathematics? Are they just frivolous? Did you learn anything from this example?

12.8 Mathematics is extremely useful as a tool for describing reality, but sometimes you have to supplement it with common sense!

With 49 soap ends, one can clearly make seven bars of soap. But these seven bars give rise to seven soap ends from which one can make another bar! So the answer is 8 bars (plus a soap end?)

TOPIC FOR INVESTIGATION: A frog is at the bottom of a 30-foot well. Every day it climbs up 3 feet, but every night it slips back 2 feet. How long does it take to climb out of the well?

12.9 Clearly, the first term 1 of the sequence is a perfect square, so let's ignore this trivial case.

All of the terms of the sequence are odd numbers so any square must be the square of an odd number, since the square of an even integer must be even.

Now, $(2n + 1)^2 = 4n^2 + 4n + 1 = 4(n^2 + n) + 1$ which when divided by 4 leaves remainder 1.

But $111\ldots111 = 111\ldots00 + 11$ and the first of these terms is divisible by $100 = 4 \cdot 25$ and so is divisible by 4, while the second term leaves remainder 3 when divided by 4. So $111\ldots11$ leaves remainder 3 when divided by 4. So $111\ldots11$ cannot be a square, since an odd square leaves remainder 1 when divided by 4. Neat huh?

TOPICS FOR INVESTIGATION: How about the sequences $2, 22, 222, 2222, \ldots, 3,$ $33, 333, 3333, \ldots,$ up to $9, 99, 999, 9999, \ldots$? Do they contain any squares apart from the obvious ones?

Can $111\ldots11$ ever be a perfect cube or a higher power?

The notion described here is called congruence due to the great mathematician Gauss (1777–1855). It is one of the most lateral and useful concepts in mathematics and you can read about it in any textbook on number theory or online.

12.10 Since the answer is negative for equilateral triangles and positive for squares, intuition gives us little information about whether it is possible to dissect a right-angled isosceles triangle into a finite number (>1) of right-angled isosceles triangles, no two of the same size. In fact, a little experimentation with a small number of pieces might convince you that the answer is negative. Certainly it is not possible with just two or three pieces.

But here is a lateral thought – if such a dissection is impossible, the proof is likely to be long and difficult, involving strange and unfamiliar techniques. Would I therefore have included it in this fun book? Very probably not! How much nicer to show a simple solution, with just six pieces, and here it is: Isn't it beautiful?

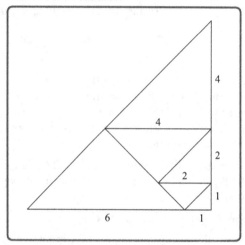

TOPICS FOR INVESTIGATION: The given dissection has six pieces. Is this the smallest possible number? Do contact me if you can show that or achieve the desired result with a smaller number and I will be very excited to hear from you and to include your discoveries in any future edition of this book. And can you think of another shape of triangle which can be dissected into n different pieces similar to itself, for $n = 2, 3, 4, 5, \ldots$?

Think laterally, think simply!

Printed in the United States
by Baker & Taylor Publisher Services